BEYOND EVOLUTION

ALSO BY DR. MICHAEL W. FOX

Concepts in Ethology: Animal Behavior and Bioethics (second edition) (1998)
Eating with Conscience: The Bioethics of Food (1997)
The Boundless Circle (1996)
Superpigs and Wondercorn: The Brave New World of Biotechnology (1992)
You Can Save the Animals: 50 Things You Can Do Now (1991)
Supercat: Raising the Perfect Feline Companion (1990)
Superdog: Raising the Perfect Canine Companion (1990)
St. Francis of Assisi, Animals, and Nature (1989)
The New Eden (1989)
Agricide: The Hidden Crisis That Affects Us All (1986)
Laboratory Animal Husbandry (1986)
Farm Animals: Husbandry, Behavior, and Veterinary Practice (1984)
The Animal Doctor's Answer Book (1984)
The Whistling Hunters: Field Studies of the Asiatic Wild Dog (Cuon alpinus) (1984)
Love Is a Happy Cat (1982)
How to be Your Pet's Best Friend (1981)
The Healing Touch (1981)
One Earth, One Mind (1980)
Returning to Eden: Animal Rights & Human Responsibility (1980)
The Soul of the Wolf (1980)
The Dog: Its Domestication and Behavior (1978)
Understanding Your Pet: Pet Care and Humane Concerns (1978)
Between Animal and Man: The Key to the Kingdom (1976)
Concepts in Ethology, Animal, and Human Behavior (1974)
Understanding Your Cat (1974)
Understanding Your Dog (1972)
Behavior of Wolves, Dogs, and Related Canids (1971)
Integrative Development of Brain and Behavior in the Dog (1971)
Canine Pediatrics (1966)
Canine Behavior (1965)

The
Genetically
Altered
Future
of Plants,
Animals,
the Earth . . .
and
Humans

BEYOND EVOLUTION

Dr. Michael W. Fox

The Lyons Press

Printed in the United States of America
Design by Cindy LaBreacht

Portions of this book appeared, in slightly different form, in *Superpigs and Wondercorn*, 1992, by Dr. Michael W. Fox.

10 9 8 7 6 5 4 3 2 1

Library of Congress Cataloging-in-Publication Data
Fox, Michael W., 1937–
        Beyond evolution: the genetically altered future of plants, animals, the earth—and humans / Michael W. Fox.
        p.   cm.
        Includes bibliographical references and index.
        ISBN 1-55821-901-3 (cloth)
        1. Genetic engineering—Moral and ethical aspects. 2. Agricultural biotech-nology—Moral and ethical aspects. 3. Genetic engineering—Social aspects. 4. Agricultural biotechnology—Social aspects. I. Title.
        TP248.6 .F687 1999
        174'.957—dc21   99-12866
                                                                                            CIP

He who would possess the world, shall lose it.

—Lao Tzu

Whenever we attempt to amend the scheme of Providence, and to interfere with the government of the world, we need to be very circumspect lest we do more harm than good.

—Benjamin Franklin

Contents

ACKNOWLEDGMENTS

First, I want to thank my wife, DEANNA KRANTZ, for her encouragement and wisdom as a sounding board for my concerns and beliefs, and Ellen Truong for editing and typing many provisional drafts of this book.

I also express my appreciation for the Lyons Press editor Chris Pavone for his insightful grasp of this complex subject and for his skills in creating a more reader-friendly book than I could have ever hoped to achieve alone.

Introduction

I am glad that the Lyons Press asked me to write a new book on the subject of genetic engineering rather than update my 1992 book *Superpigs and Wondercorn*. There have been so many developments since 1992 that in order to do justice to the topic, a new book is in order—but not one packed with references to research on the creation of new crop varieties or new kinds of genetically engineered mice, of which there are now thousands. There has indeed been a "bioexplosion" in laboratories around the world, which are busily sequencing, identifying, isolating, and switching genes among all kinds of species, from putting scorpion-venom genes into various crops to splicing human genes into mice, pigs, and goats. To reference all these "new creations," many of which have been patented, would make tedious reading. So what I have done in this book is to select examples that show what genetic engineers are doing to animals, plants, and other living beings such as bacteria and viruses. The emphasis is not on the patented *hows* of this new technology, but on the *whys* and on the *consequences* that take us beyond natural evolution into a genetically altered future.

Many of the predictions on "the brave new world of biotechnology and where it all may lead" in *Superpigs and Wondercorn* a mere seven years ago have come true. Nonetheless, some critics panned that book, saying it was full of doom and gloom and did not give a balanced view of biotechnology's promises and emerging benefits. I was quite clear in stating that some of the fruits would be medical and forensic benefits, from DNA typing and diagnostic probes to new and more effective pharmaceuticals, and also industrial oils, biodegradable plastics, and other new bioindustrial products such as enzyme cleaners and detoxicants. But as with any technology, the main issue is how and to what ends it will be used, and I predicted that the use of biotechnology in agriculture and biopharming (using genetically engineered animals to produce various pharmaceuticals) would raise a host of ethical, social, and environmental problems if not carefully applied and effectively regulated.

The direction being taken and the whys of biotechnology in agriculture should concern us all, because we consumers are now the guinea pigs for testing new genetically engineered foods. An estimated 60 percent of processed foods now contain some genetically engineered ingredients. Why should this concern us? Because the very nature of genetic engineering is fraught with unforeseen consequences, and because the pre-cautionary principle is not being applied by the life-science industry for reasons of cost and limited scientific ability to pre-dict consequences.

Splicing genes from one organism into another will produce changes in the complex composition of the new genetically engineered life form. When biotechnologists focus only on introducing a desired trait, such as herbicide resistance in soy-beans, for example, other unexpected changes in the plant's composition may be overlooked. These changes may come to light after more research, but because of costs, such research would be undertaken only if the new life form causes recog-nized consumer health or environmental problems.

Harm that is being done to farm and laboratory animals should also concern us as humanitarians and as consumers (if we are not vegetarians). Already, harm and risk are accepted by the biotechnology industry such that, in its quest to commodi-tize life, animals are being deliberately created with genetic defects that will guarantee that they suffer sometime after they are born. Thousands of varieties of mice with genetically engi-neered diseases have been created, further imbedding in the public mind the belief that this is medical progress and the good ends justify the evil means of deliberately creating animals that will suffer. Plants are deliberately engineered to produce toxins that kill or reduce the fertility and vitality of beneficial insects. The food industry sees organic farming as a major enemy and touts the benefits of genetically enhanced, ostensibly more nutritious crops instead of first enhancing the soil organically to make crops naturally healthy and nutritious. Thousands of dairy cows have been put at risk by milk producers who inject them with an engineered hormone (rBGH) that supposedly will make them more "efficient" and "productive"; many of these cows break down physically, become diseased and nonproductive, and are slaughtered in their prime.

I will elaborate on these and other harms and serious risks, not to foster doom and gloom or "biotechnophobia," but to present for public review what the architects of a new world order based on genetic-engineering biotechnology are doing— and how it will affect us now and forever.

Concerns about genetic pollution, loss of wildlife and biodiversity, and disruption of ecological and evolutionary processes will also be discussed in this book. Concerns and conclusions are supported with documented research reports and expert opinion.

I have included many reference resources and key organizations for readers to join to keep up to date on developments in biotechnology that will affect our lives for generations to come. More public involvement and better understanding are essential in helping ensure that these developments, especially in agriculture, do more good than harm. This is particularly important now that the U.S. government, as many others, has virtually abdicated its oversight and regulatory responsibilities in the name of progress for a U.S. industry facing a highly competitive world market.

My major concern, which I hope readers will thoroughly consider, is the state of mind or worldview that is behind the development of a new industry and world order based on genetic manipulation, control, and monopoly. Most of these developments in genetic commerce have been produced in a virtual ethical vacuum, often in secrecy. And as many critics point out, government oversight has been minimal and public involvement precluded.

Human civilizations through the ages can be characterized by a particular worldview or state of mind that is an amalgam of various beliefs, values, and aspirations. The worldview of biotechnology and the new "life-science" industry that it has birthed is an outgrowth of the dominant worldview of industrial civilization and is not, contrary to its claims, based on either sound science or objective reason.

This is the central concern, therefore, of this book, because in the wrong hands, and with this dominant worldview driving biotechnology, our newfound power over the genes of life will do more harm than good, and in cause and effect will amount to a new world disorder.

When the ability to crack and rearrange the genetic code, imprinted into the double helix, was discovered, the paired spiraling molecules of DNA became the next frontier of nature to be exploited. The new life-science industry, served by a scientific priesthood of genetic engineers, is changing what some theologians call God's words, encoded in the DNA of all living beings, into their own lexicon of useful and profitable market products.

Applied with a very different worldview, one of great caution and humility, which sees every creature and creation as the word of God, this newfound power over creation and evolution could produce great good. There must be reason within reason as well, knowing that we do not know enough to guarantee that genetic engineering will have no harmful consequences.

Many people have become aware recently of the social and ethical problems associated with changing the human genome by genetic engineering, and with the specter of cloning people. But these issues are not the most serious ones in biotechnology at present, hence, this book does not cover them. The same bioethical constraints that will help maximize the medical benefits of biotechnology are those that most urgently need to be applied in the now global expansion of producing genetically engineered crops and foods.

We cannot yet create life or stop death forever with biotechnology, but we can and do control and alter much of life on earth already to serve our own pecuniary ends. The biological, ecological, evolutionary, ethical, and social consequences of genetically altering life cannot be ignored.

Will this new technology mean the end of the natural world as the human species creates a new world order beyond evolution? Genetic engineering raises many questions such as this that should not be ignored.

How do you as a consumer feel about new genetically engineered crops that produce their own pesticides, and "Franken foods" made from various genetically engineered ingredients now being marketed? What about the new varieties of transgenic farm animals that I call "humanimals" because they have been engineered to have human genes so that their bodies produce valuable biopharmaceuticals in their milk? What about the risks of "genetic pollution," by which genetically engineered organisms (GEOs) released into the environment transfer their genes to closely related species? (This has actually already happened.) What about the twenty thousand and more varieties of transgenic mice that have been created to serve as models for human diseases and for testing new drugs? Is this really cost-effective and scientifically sound medical progress, or is it another intensification of animal suffering primarily for profit? Will the organs of transgenic pigs that carry human genes to make them more "donor friendly" to human recipients—soon to be marketed—be safe?

It may seem like an overstatement to assert that the integrity and future of creation is threatened more by this new technology than by any other past human invention or activity, including nuclear fission and the development and release of petrochemicals into the environment. But by the end of this book I believe that the evidence presented will have convinced even those who are enchanted by the promises of genetic-engineering biotechnology that it is already dangerously on the wrong path.

There is a right path. Critics who dismiss these concerns are living in denial. Those responsible for the creation of altered life must also be responsible for the future integrity of the natural world, for public health and safety, and for the socially just and equitable use of life's genetic resources.

What future wonderland is the genetic sorcerer creating for us? I foresee virtual-reality zoological parks with cloned and forever preserved endangered species—pandas, Siberian tigers,

Nilgiri langurs—but never enough species to replenish and restore natural ecosystems. Is the death of nature, even the end of natural evolution, inevitable? Shall we accept the expropriation of God's creation, the disenfranchisement of an omnipresent divinity that is inside all things? We have become blind to the perfection of larks, locusts, and loco weeds within healthy ecosystems in a once perpetual state of regeneration and transformation. We harm and destroy myriad interconnected species, co-evolving, co-creating, maintaining, and ever changing forest, jungle, savanna, lake, and ocean ecosystems. With each extinction, a stargate closes, and a connection in the once seamless web of planetary life is severed forever.

Many extinctions—genocide indeed and ecocide—are justified on the grounds of "progress" and profits, which escalate ecological imbalances. There are new weeds, plagues, and pestilences against which we wage war without understanding that they are symptoms of ecological imbalance most often caused by us. For example, we choose to raise crops in ecologically unsound ways, using synthetic herbicides and chemical fertilizers that sterilize the living soil and make plants sick and more prone to blights and pests. Because we choose to incarcerate animals in factory farms and feedlots, we release a Pandora's box of new zoonotic bacterial diseases that cause food poisoning and death in consumers of meat, eggs, poultry, dairy products, and seafoods.

When applied to biological systems such as agriculture to increase productivity at all costs, industrial science can have disastrous consequences. Now industrial science is using the new tools of biotechnology to increase crop and farm animal productivity, which will only worsen the uncorrected problems of production agriculture.

Intensive industrial agriculture is adopting genetic engineering like gangbusters because the sellers of genetically engineered organisms and products now control much of the food industry. Intensive petrochemical-based agriculture has been ecologically and socially devastating.[1] Building upon that agrichemical foundation, agribiotechnology will mean a continuation of ecologically and socially harmful, world market–driven farming practices and agricultural policies, putting consumers' health even more at risk in the process.

These issues should concern and involve us all. We must ensure that our power over the genes of life and the future of creation is tempered by reason and compassion to improve the human condition and enhance the life and beauty of the natural world.

CHAPTER 1

In the Beginning

In the beginning was the Word,
and the Word was with God,
and the Word was God.
                    —John 1:1

Since genetic engineers have the power to change the future of all life on earth profoundly and forever, we may better foresee the outcome of their influence on creation or biological evolution if we understand their worldview, or "paradigm." So this chapter examines the ideological and historical foundations of the life-science industry of biotechnology—foundations, I believe, that are not sufficient to guarantee the safe and wise use of genetic engineering.

After the Copernican revolution around 1500 A.D., science gradually became the new gospel. The paradigm of modern science emerged subsequently in Europe during the so-called Enlightenment period of the seventeenth and eighteenth centuries. It was such a human-centered view that there were no

moral or ethical boundaries to constrain and contain our
actions, usually harmful, to other animals and nature, unless the
animals and the land were someone's property. We can trace this
worldview of human mastery and dominion over creation, along
with ethical blindness and instrumental rationalism, from one
civilization to the next, from the Greek and Roman imperial
empires to the pre- and post-industrial colonial empires and into
the new age of global economic imperialism.

Enlightenment philosophers such as René Descartes and
Francis Bacon based their new worldviews, which helped form
the basis for modern science and the industrial revolution, on
the Catholic and Protestant legacy of believing that only
humans were made in the image of God and have immortal
souls, and that animals were created for man's use. They
embraced the dualistic views of the separateness of humans
from animals, matter from spirit, mind from body, man from
nature, and nature from God.[1]

Bacon urged that we "vex Nature of her secrets," have
"power over her and improve upon her," and "have commerce
with her"; he gave religious sanction to man's "endeavor to
establish and extend the power and dominion of the human
race itself over the universe." He actually wrote a rare piece of
fiction, a utopian saga called *New Atlantis*, in which there were
bizarre human-engineered animals, as in the more recent tale of
The Island of Dr. Moreau. He described animal parks and enclo-
sures that were used not only for public viewing but also "for
dissection and trials, that thereby we may take light what may

be wrought upon the body of man. . . . We try all poisons and other medicines upon them as well as of chirurgy and physic."[2]

Descartes maintained that only humans can reason and that animals are unfeeling machines. He even condoned the first biomedical experiments on animals, mainly dogs and cats, which were done without any anesthesia. He dismissed the screams of tortured dogs, convinced that animals have no real feelings of which they are aware, because to be aware means to have a conscious self, and animals are not self-conscious. He contended that the screams were nothing more than the breaking down of the machinery of the dog's body.

After thirty-five years of experience as a veterinarian concerned about how humans regard and treat animals, I submit that modern science is built on this anthropocentric and mechanistic worldview of the age of Enlightenment.

VIEWS OF THE SCIENTIFIC COMMUNITY

I have never felt more alienated from my own kind than when, in the spring of 1985, I confronted the National Institutes of Health Genetic Engineering Committee in Washington, D.C. Jeremy Rifkin of the Foundation on Economic Trends and I challenged the National Institutes of Health to temporarily suspend government-funded transgenic animal research until the ethics and consequence of developing new industries out of this biotechnology had been fully explored and publicly aired. We were met with united opposition. In the committee's large conference room—with scientists gathered around a thirty-foot-long oval table and the press and observers seated around them—I experienced a sense of vertigo and unreality as the chairman read statements from scientists supporting transgenic research. These statements came in rebuttal to the ethical question that Rifkin and I had raised about the rightness of interfering so profoundly with the *telos*, or inherent nature, of animals through such manipulation. One statement implied that this was a perfectly natural development in human evolution: to play God. Another insisted that animals have no inherent nature (i.e., no intrinsic worth) because their *telos*—or final end—is death. Thus it was reasoned that there was nothing wrong with directing that final purpose to satisfy purely human ends.

That none of the academically esteemed scientists and bioethicists on the committee questioned these assumptions and this wholly instrumental and self-serving attitude toward life was the most shocking experience I have had in my career as a scientist and spokesperson for animal welfare and rights.

Here is a selection of statements by various scientists and physicians that were handed out at this unforgettable meeting:

> The idea that a species has a "telos" is contrary to any evidence provided by biology and belongs rather in the realm of mysticism. That mysticism is a poor basis for sound public policy is amply confirmed by history. (Professor M. J. Osborn, chairman, Department of Microbiology, School of Medicine, University of Connecticut)
>
> History, from Galileo through Lysenko, teaches us that mysticism can never yield rational and wise public policy in scientific matters. . . . The notion that a species has a telos (a

purpose) contravenes everything we know about biology.
Species can have, and many in the past have had, a telos (an
end), namely, extinction. That is the only telos known to exist.
(Dr. Maxine Singer, National Institutes of Health)

These scientists dismiss philosophy and ethics as mysticism,
failing to grasp the full and original meaning of *telos*.* But the
debate about telos is a matter of semantics. The real issue is
whether living things have inherent natural qualities that we
tamper with at our peril. I believe that they do. If this is mysticism, so be it.

Another biomedical scientist who wrote to the National
Institutes of Health in support of transgenic research presented
an evolutionist's rationale for transgenic manipulation—a kind
of biological determinism:

> We humans are participating in the process of evolution *per se*.
> By that I mean that our ability, acquired through evolution, to
> manipulate genomes by selective breeding and more recently
> by recombinant DNA technology is an integral component of

---

*Aristotle recognized that form and matter coexist and contended that the
"cause" of any changes in these can be subdivided into four parts: (1) formal
cause, which is the essence or pattern in matter that persists as the matter under-
goes change; (2) material cause, which is the substratum, the matter in which the
essence or form is contained; (3) efficient cause, which refers to the proximate
agents of change; and (4) final cause, or the purpose for which the change was ini-
tiated. *Telos*, defined in *Webster's Third International Dictionary* as "an ultimate
end or object," includes all four of these subdivisions, a point often overlooked by
those who regard telos as referring only to final cause. Since telos also relates to the
qualities of an object (formal cause and material cause), it is correct, as philosopher Bernard Rollin has insisted, to infer that the term *telos* refers to an inherent
nature or beingness and not simply to the ultimate end or final purpose of an
object of living entity.

evolution itself and is not, as has been claimed in the past, "tinkering with evolution." Instead, *it is* evolution. Because of our evolved level of consciousness, we as human beings must, however, utilize appropriately our consciousness and ability to anticipate as we effect our role in the evolution of our own and other species on this planet. Thus, I do not subscribe to the premise that we as human beings must through regulatory agencies exercise control over evolution by forbidding specified acts of nature, we being mere agents of nature. (David W. Martin, Jr., M.D., School of Medicine, University of California, San Francisco)

Yet other supporters of transgenic research argued that it is ethically acceptable because we have been crossbreeding and selectively breeding animals for centuries (ignoring the data showing that such manipulations have increased animal disease and suffering through genetically related and "domestogenic" diseases). For instance, Professor E. Brad Tompson, chairman of the Department of Human Biological Chemistry and Genetics at the University of Texas Medical Branch at Galveston, wrote:

The truth is that man has been experimenting with crossing animal species since time immemorial. The technology available to do it now simply differs from that available formerly. It is, in my opinion, dangerous and wrong for a prohibition of the sort suggested to be put into place as part of the framework in which American research is conducted. It would undoubtedly deter important and potentially useful experiments from being done, experiments which would have potential for improving the lot of many species including but not limited to mankind.

Some scientists seem to believe that there is no real difference between the traditional techniques of selectively breeding domestic animals for certain useful traits and the new engineering technique of inserting "useful" genes of other species. The latter is seen simply as an extension of the former. For example, Dr. Cornelius Van Dop of the Johns Hopkins Hospital, Baltimore, wrote:

> The selective breeding of animals directed to amplifying or eliminating certain traits has been a human activity since the first mammal was domesticated during prehistoric times. This selection for specific traits (mutated genes) has irreversibly modified the gene pools of innumerable species for man's economic gain and whim. . . . Current bioengineering technology stands at the threshold of being able to selectively modify one gene at a time and thereby reduce dependence on selective breeding for altering certain traits. The selective introduction of foreign genes into germ lines is thus a logical extension of animal husbandry.

There is a world of difference between genetic engineering and selective breeding. Transgenic manipulation entails crossing the natural biological boundaries *between* animal species. This has *not* been done before. (The closest analogy is the crossbreeding of closely related subspecies such as horses and donkeys or wolves and dogs.) Single genes can have profound consequences, and increasing the utility of pigs with cattle growth genes, or cattle with elephant growth genes, could so disrupt the animals' telos (intrinsic nature) biophysically, metabolically, and developmentally as to create a host of health and welfare problems that would require further technological "fixes" with chemicals, drugs, and additional genetic manipulation.

Jeremy Rifkin observes that:

> Already researchers in the field of molecular biology are arguing that there is nothing particularly sacred about the concept of a species. As they see it, the important unit of life is no longer the organism, but rather the gene. They increasingly view life from the vantage point of the chemical composition at the genetic level. From this reductionist perspective, life is merely the aggregate representation of the chemicals that give rise to it and therefore they see no ethical problem whatsoever in transferring one, five or a hundred genes from one species into the hereditary blueprint of another species. For they truly believe that they are only transferring chemicals coded in the genes and not anything unique to a specific animal. By this kind of reasoning, all of life becomes desacralized. All of life becomes reduced to a chemical level and becomes available for manipulation.*

Dr David Baltimore, former director of the Whitehead Institute for Biomedical Research, Cambridge, Massachusetts, expressed the technocrat's instrumental ideology of separating science from ethics and morality (because the latter are subjec-

---

*J. Rifkin, *Algeny: A New Word—A New World* (New York: Viking, 1983), p. 47.

tive), and of leaving ethical decisions, such as transgenic manipulation, to rationalism:

> I oppose "prohibitions" on the grounds that they provide an apparent simplicity that often leads to difficulty. I also oppose writing into regulations statements about "morally and ethically unacceptable" practices because those are subjective grounds and therefore provide no basis for discussion. There are good scientific grounds for not putting any new genes into the human gene line today and I believe that we should rest our behavior on such rational assessments not on the shifting and personal grounds of morality.

It is ironic that most scientists and select committees believe that although there is no ethical issue involved in switching genes between animal species and putting human genes into animals, it is wrong for animal genes to be put into human gene lines. Clearly, to them, only human life is sacred.

Other letters from biomedical scientists emphasized that transgenic research is essential for understanding how genes work and can be controlled so that genetic and developmental defects in children might be corrected. But they miss the point that medical genetic engineering is interventive rather than preventive, and since genetic and developmental defects are in part environmentally correlated (e.g., with teratogenic and mutagenic agrichemicals, food additives, industrial chemicals, and pollutants), genetic engineering will be no panacea if environmental factors continue to be ignored for political and economic reasons.

SOME SCIENTISTS' ATTITUDES TOWARD NONHUMAN LIFE

My thesis is that since the Enlightenment, a belief and knowledge system—along with industrialism, materialism, and consumerism—has evolved and is now at the controls in genetic-engineering laboratories around the world. This belief system is defective, with a potentially fatal flaw.[3] Its worldview cuts off empathy, feeling, and therefore respect and concern for other sentient beings. How else could the biomedical priesthood condone incarcerating our simian tree-swinging monkey cousins from the forests and jungles in cages three feet wide by four feet long and three to five feet high?

In the 1980s, I caused a ruckus at the National Institutes of Health (NIH) and other animal laboratory facilities when I sent letters to several veterinarians in charge of animals being used in biomedical research asking what analgesics they used, how much of them, and when, in what species, and for how long. Many, I discovered, were using none to help alleviate post-operative pain following various experimental procedures. I learned later that several meetings were held to discuss the topic and set up some guidelines.

The question of post-operative pain in these vivisected animals had evidently never been considered. Was it because of a lack of empathy or a belief that animals don't really suffer? Or was it more a question of denial?[4]

Before an audience of pig farmers in Des Moines, Iowa, in 1985, I asked an animal scientist and advocate of factory farming with whom I often debate, "Well, Stanley, do you think pigs have feelings?" He replied, "We need to do more research before we can be really sure." On another occasion around the same time, at a congressional hearing where I testified in support of a bill to banish veal crates, Stanley said, "There is no scientific evidence that veal calves need to turn around."

To make a pronouncement such as this, to think in such a way—or was it to protect the factory veal industry? I wonder what kind of feeling and therefore vision exists in such a person, who is nevertheless of high academic rank. And he is not alone in his views.

In a recent survey of academics from various disciplines, a substantial number of those in animal science and zoology (17 to 25 percent) did not believe that animals have minds. From 67 to 100 percent of all participants said they perceived that animals have the ability to think, but a substantial number (6 to 33 percent) of animal scientists, zoologists, veterinarians, and English faculty said no, animals don't think.[5]

We have now examined the feeling state and the vision of "establishment" biomedical scientists, some of whom are at the helm of creation, able to artificially conceive, clone, and genetically transform all manner of species, human and nonhuman. They can switch genes on and off and replace different body parts,[6] and soon possibly resurrect extinct species from long-frozen DNA. Is their worldview adequate, even appropriate, as a foundation for genetic-engineering biotechnology? Or has this technology advanced too fast for our ethical sensibility, moral imagination, and scientific knowledge to enable us to maximize its benefits and minimize its risks?

This new technology was born in 1953 with the first step in breaking the genetic code, as a result of the serendipitous discovery of two British scientists, James Watson and Francis Crick, coupled with the collective knowledge base of molecular biology. Watson and Crick demonstrated that nucleotides, which are the building blocks of deoxyribonucleic acid (DNA), are arranged in a double spiral, or helix. They were subsequently awarded the Nobel Prize for this discovery. Few would have thought that a multibillion-dollar-a-year industry would result as scientists developed ways to alter this double helix by cutting out portions and splicing in segments containing genes from other species.

The sequence of nucleotides that determines which proteins will be made is commonly called the genetic code. Distinct differences in the sequence produce a different genetic code for every species and individual variant on earth. But all life forms share the same basic genetic structure: the double helix of DNA. The stepping stones that paved the way to today's global genetic-engineering industry after Watson and Crick's breakthrough were quickly put in place. In 1973 Stanley Cohen of Stanford University and Herbert Boyer of the University of California at San Francisco were the first to splice recombinant DNA into bacteria, which then multiplied (by natural cloning), producing copies of the foreign genetic material. This technique was patented and soon resulted in the development of genetic-engineered bacteria to produce a variety of profitable products, from analog calves' rennet (for cheese making) to analog human insulin (for diabetics).

In 1977 a revolutionary technique called the polymerase chain reaction was developed and patented by scientists in the United States and the United Kingdom, which enabled the rapid sequencing or reading of the order of nucleotides in DNA molecules.

A human ear cannot be grafted onto a mouse because it will be rejected, unless the mouse's immune system has been knocked out. (This has been done to produce a mouse that carries a whole human ear!) The genetic barrier between different species, which in nature permits almost no genetic exchange, except at the simple cell level of bacteria and viruses, is probably there for evolutionary and other good biological reasons. This new technology enables genetic engineers to break through that barrier between species.

Human genes and other foreign genes can now be put into mice, pigs, tomatoes, and other very different life forms and not be rejected. Even though they aren't rejected, however, foreign genes can cause harm to their recipients. They may be "overexpressed," producing too much of a particular protein, such as a growth hormone, for example, and result in abnormal growth and crippling. Foreign genes that create different proteins and convey various hereditary traits can be more and more easily switched between species as the techniques of "gene splicing" continue to be improved by genetic engineers.

Some viruses that invade plant and animal cells have been engineered to carry spliced-in genetic material to create transgenic plants and animals. Another technique is to use a "bioblaster," a small particle accelerator that drives microscopic particles of the inert metal gold, covered in genetic material, into the cells of seeds and live animals in the hope of creating a new and profitable life form. Yet another technique employs microsurgery, implanting certain genes on segments of DNA that have been cut up at precise locations by what are called restriction enzymes, which act like scissors.

By 1988, the dream of deciphering the genetic sequences in humans and other animals had become a reality, and the multibillion-dollar (public-funded) Human Genome Project was launched by the National Institutes of Health under the direction of James Watson. By means of new super-computer technology, a worldwide network of collaborating laboratories was quickly established, along with media hype that featured phrases such as "the end of disease" and "the promise of human perfection."

CORPORATE INVESTMENT AND DEVELOPMENT

The first corporation to commercially develop this new technology, Genentech in San Francisco, was established in 1976. Its initial public offering in 1980 set a Wall Street record for the fastest price-per-share increase ($35 to $89 in twenty minutes). A year later, another biotech company, Cetus, set a record for the largest amount of money raised in an initial public offering ($115 million). Over eighty new biotech firms had been formed by the end of 1981. Such was the excitement of venture capitalists over the promising beginnings of this new technology.

The Technology Transfer Act was passed in 1986, providing expanded rights for companies to commercialize government-

(i.e., public-) funded research. During these formative years, the U.S. biotechnology industry, which survived the October 19, 1987, crash on Wall Street, faced intensifying competition from other countries venturing into this new field, notably Japan, Germany, the United Kingdom, and the Netherlands. Market confidence returned, however, after many smaller U.S. biotech companies had either gone under or been bought up by larger petrochemical and pharmaceutical multinationals. The industry sold $17.7 billion in new stock in 1991, another Wall Street record.

In 1980, amid great controversy and public opposition from many religious leaders and bioethicists, a U.S. Superior Court ruled (by a 5-to-4 vote) that an oil-eating genetically engineered bacterium developed by General Electric Company could be patented, overturning a policy of the Patent and Trademark Office that life could not be patented.

Also in 1980, scientists at Ohio University in Athens announced that they had succeeded in creating the first transgenic mice, which grew twice as big twice as fast after they had been "spliced" during early embryonic development with human growth genes.

In 1985, the U.S. government permitted the first deliberate release of genetically engineered organisms into the environment: Omaha's Biologics Corporation's genetically engineered swine vaccine, and California's Advanced Genetic Sciences, Inc.'s genetically engineered bacterium (Frost-ban), which is sprayed on strawberries to act as an antifreeze. These releases were subsequently protested by environmentalists and other informed citizens.

The creators of the first animal ever to be patented, the "Harvard oncomouse," a mouse genetically engineered to be highly prone to breast cancer, had their application approved in 1988 by the U.S. Patent and Trademark Office soon after that office, under industry pressure and in spite of public protest, announced that nonhuman animals are patentable subject matter. In 1990, with virtually no public comment, the FDA approved genetically engineered renin, an enzyme used to produce cheese, opening the door for other genetically engineered food additives. The following year, the Environmental Protection Agency (EPA) approved the first genetically engineered biopesticide for use in agriculture.

RECENT DEVELOPMENTS—AND CONTROVERSIES

The breathtaking pace of research and development in biotechnology during the last decade of this millennium parallels an increase in public concern and growing opposition to the creation and release of genetically engineered organisms into the environment. The U.S. Department of Agriculture's Animal, Plant and Health Inspection Service (APHIS) permitted field trials of seventeen new varieties of genetically engineered crops in July 1994 after finding that they would have "no significant environmental impact." According to informed ecologists, that finding had no sound scientific basis.

The U.K. Advisory Committee on Novel Foods and Processes was more cautious, calling for removal of antibiotic-resistance markers (used to identify genetically engineered plants and foods) in uncooked, "live" foods (yogurt and baker's yeast, for example) and urging the removal of such markers, widely used by the biotech industry, in a variety of other crops. The concern was over the risk of transfer of antibiotic-resistant genes to bacteria in the human gut. (See p. 153 for more about this.)

The U.S. Food and Drug Agency, which has jurisdiction over food safety, declared in November 1994 that seven new genetically engineered (GE) crops and foods were safe for consumers, including virus-resistant squash, herbicide-tolerant soybeans, and pesticide-producing "Bt" potatoes, tomatoes, canola, and cotton. The FDA concluded that there was no need to label these foods as genetically engineered since labeling requirements should depend on safety and nutritional characteristics, not on the method of development. Yet these products contain antibiotic markers that are considered a public health risk in the U.K.

The U.S. Environmental Protection Agency also has regulatory authority over the safety of genetically engineered crops that produce their own pesticides. This regulatory oversight is not integrated with the FDA's responsibility for food safety, however, or with the USDA's APHIS involvement in plant and animal health. Without any integrated regulatory program, legitimate environmental and consumer health concerns often fall through the cracks. An inquiry to the USDA about the health risks of Bt corn, for example, would be referred to the FDA and from there to the EPA because only the EPA has authority over pesticides. In March 1995 the EPA approved for the first time the planting of GE crops of cotton, potatoes, and corn that produce an insect-killing toxin from *Bacillus thuringiensis* (Bt). This approval was opposed by organic farming, consumer health, and environmental public-interest organizations. By 1997 several Bt-producing crops, such as apples, rice, tomatoes, walnuts, cotton, tobacco, and potatoes, had been created—and EPA-approved—even though Bt is known to kill many beneficial insects, and harmful insects can quickly develop resistance. Organic farmers, who have used Bt for thirty-five years as a natural alternative to synthetic pesticides, now face the specter of Bt-resistant insects, which these GE crops will eventually create. A coalition of organic farmers and farming associations filed suit against the EPA in September 1997, alleging that in approving transgenic plants that express Bt toxin, the EPA was seriously threatening the future of sustainable agriculture.

The role of the U.S. government in promoting the interests of the biotechnology industry and the lack of government concern over the potential risks of GE crops were evident at a meeting I attended in 1995 at the United Nations headquarters in New York. Several nongovernmental organizations from around the world had managed to put the issue of the risks of GE crops on the agenda of the Biodiversity Convention (which the United States had not yet joined). But the U.S. government obstructed attempts at the United Nations Commission on Sustainable Development to establish binding international biosafety and bioethical protocols for GE plants, other organisms, and GE food.

Later that year at the International Biodiversity Seminar in London, HRH Prince Charles shocked biotech advocates when he asked, "Am I *really* alone in feeling profoundly apprehensive about many of the early signals from this brave new world [of biotechnology] and the confidence . . . bordering on arrogance . . . with which it is promoted?"

Adding fuel to the growing public unrest and scientific uncertainties about biotechnology, in September 1995 some EPA staff members published a report entitled *Genetic Genie: The Premature Commercial Release of Genetically Engineered Bacteria*, charging their own organization with failure to adequately assess risks and being over eager to promote biotechnology.[7]

Controversies continued to escalate both nationally and internationally as new developments in genetic engineering became known to the public, and public-interest organizations more closely monitored corporate and government activities. In January 1996, the U.S. Congress cut the funds of and closed down the USDA's Agricultural Biotechnology Research Advisory Committee and the Office of Agricultural Biotechnology (OAB). Critics interpreted this as further deregulation, as implying that the organizations' "mission of implementing and coordinating the USDA's policies and procedures concerning agricultural biotechnology" had been completed. The OAB director, Dr. Alvin Young, was moved to the Foreign Agriculture Service to help carry "the torch forward to help settle agbiotech issues worldwide."[8]

Another controversy erupted over the inclusion of genetically engineered products under organic farming standards being put together by the USDA, which the biotech industry obviously wanted. In November 1996, the National Organic

Standards Board, a newly constituted advisory group under the auspices of the Agricultural Marketing Service of the USDA, in concert with the European International Federation of Organic Farming Movements, unanimously agreed that *no* transgenic produce be eligible for classification as "organic." This position was subsequently adopted by the joint Food and Agriculture Organization and World Health Organization's international food safety codes, called the Codex Alimentarius.

The results of two public surveys released in April 1997, one by the International Food Information Council and the other by biotech and pharmaceutical conglomerate Novartis, found that 78 percent and 93 percent of U.S. consumers, respectively, want genetically modified foods labeled as such. The Novartis survey also found that over half prefer organic produce and over half believed foods produced with pesticides to be unsafe.

A public protest regarding genetically engineered foods was held in October 1997 before the Federal Building in Seattle, protesting the Food and Drug Administration's refusal to label GE foods as genetically modified and the U.S. Department of Agriculture's stated intent to allow GE foods and crops to be certified "organic" under its National Organic Standards. In April 1998, after 280,000 letters were received protesting the USDA's attempt to allow genetically engineered crops under the National Organic Standards, the life-science corporation Monsanto requested that the USDA delay its decision for at least three years.

An unprecedented coalition of scientists, religious leaders, health professionals, consumers, and chefs filed suit in May

1998 against the FDA to obtain mandatory safety testing and pertinent labeling of all genetically engineered foods. The suit, filed in federal district court, alleged that current FDA policy, which permits such altered foods to be marketed without any testing and without being so labeled, violates the agency's statutory mandate to protect public health and provide consumers with relevant information about the foods they eat. The suit also alleged that the policy is a violation of religious freedom (GE foods containing animal genes could violate dietary ethics of Jews, Moslems, and Hindus, for example). The suit was coordinated by the Alliance for Bio-Integrity, with key collaboration from the International Center for Technology Assessment. Both are nonprofit organizations dedicated to advancing human and environmental health through sustainable agriculture and safe technologies. In the suit, plaintiffs challenged the marketing of thirty-three different genetically engineered whole foods that are currently being sold without adequate labeling or safety testing. These include potatoes, tomatoes, soy, corn, squash, and many other fruits and vegetables to which a variety of new genes from different species have been added. These genetically engineered whole foods are also used as ingredients in processed foods and are estimated to be present in 60 percent of processed food products, from major soy-based baby formulas to some of the most popular corn chip brands. Because of FDA's failure to require labeling that discloses genetic engineering, millions of American infants, children, and adults are now consuming genetically engineered food products each day without their knowledge.

The life-science industry is now poised to create a new world order—or disorder—the scope and consequences of which will be explored in the next chapter.

CHAPTER 2

Genetic Imperialism: The New World Order of Biotechnology

I've come to believe that the potential
power of genetic engineering dwarfs that
of nuclear power.
                    —Professor Liebe Cavalieri

Welcome to the Genetic Age. Genetic-engineering biotech-nology is the fastest-growing industry in recent history, and its costs and consequences are not well understood by the general public.

Biotechnologists are developing a variety of genetic-engineering techniques to produce a host of new products and life forms. There is the promise of cures for various diseases, such as alcoholism, cystic fibrosis, and kidney disease. Pigs are being bioengineered to produce human hemoglobin so they can serve as human blood donors; other pigs have been given human-like immune systems so they can serve as organ donors for people needing new kidneys and hearts. Crops are being

engineered to produce not only insecticides but also edible vaccines for people. The genes of venom-producing spiders are
being isolated and spliced into edible plants to help combat
insect pests, many of which have become resistant to chemical
pesticides. Will this new technology help feed the hungry world
and provide even drought-resistant and more nutritious crops?
Or is the development of such super crops part of the complex
of forces that actually contribute to world hunger and poverty?

Without careful deliberation and public involvement in the
decision-making process of how this new technology is applied
and its consequences, the very integrity and future of creation
may be irreparably disrupted, at great cost to future generations. We have acquired a godlike power over earth's creation
via our dominion over the genes of life. If we do not use this
newfound power to heal the earth by healing ourselves, and
heal ourselves by healing the earth, then the suffering of
humanity and the desecration of the environment will continue. Are we opening Pandora's box, releasing forces about
which we know little, whose cumulative domino effect could be
socially, economically, and environmentally devastating?

This technology is being developed by the industrial complex of petrochemical food and pharmaceutical life science. We
face a rising biotechnocracy that historically has had no respect
for the earth or concern for future generations. It has grown and
prospered from the production of pesticides, regardless of environmental and public health concerns, and it has profited roy-
ally from treating the infants and adults who often sicken as a
consequence.

New genetically engineered "wonder drugs," derived in part from tests on transgenic animals, such as those now being touted to "attack cancer at its genetic roots," give false hope to the public and distract public focus from the social, economic, and environmental causes of many diseases and their prevention. The new medicine of genetic-engineering biotechnology is highly profitable. Preventive and holistic medicines, however, threaten corporate profits and the biomedical industry. The industrial system profits from making people ill and then profits again by offering new diagnostic and treatment procedures to those in need. For example, the chemical and biotechnology multinational DuPont Nemours is one of the biggest U.S. distributors of pesticides and other cancer-causing chemicals. This same company put up the money to develop the genetic-engineering bioblaster and the Harvard oncomouse, discussed in chapter 1. This new variety of mouse is sold to testing labs to screen harmful chemicals and test for new cancer cures. DuPont also sells a high-resolution x-ray film for screening women for breast cancer, thus profiting further from the consequences of people, women especially, being exposed to an increasingly toxic environment and pesticide-contaminated food and drinking water.

As for the claim that agribiotechnology will feed the hungry world by boosting industrial agriculture's productivity: the primary emphasis has been on getting third-world farmers to use their land not to feed their communities and their many starving animals better, but to produce cash crops for *export*. More and more people go hungry without local food of good quality or the natural products they once gathered from uncultivated wild lands, which contributed to the sustainable social economy of hundreds of thousands of villages and tribal settlements around the world. Megafarms and plantations are encroaching upon and obliterating remaining wildlands from California to India, from Tanzania to the Amazon. Agribiotechnology is another nail in the coffin of indigenous peoples, supplanting traditional sustainable agriculture, aquaculture, and social forestry. In order to feed themselves and what livestock they have, those tribal gatherer-hunters, pastoralists, and village farmers around the world who do not emigrate or drift to the cities are forced into marginal and wild lands, with highly destructive consequences that include overgrazing, deforestation, soil erosion, and desertification.

The new varieties of transgenic super seeds that farmers are planting today are part of the increasingly integrated life-science industry that has grown from the poisoned roots of petrochemical-based agribusiness. The term *farmer* is a misnomer for those who produce essentially biomass vegetable oils, starches, and amino acids for various branches of the life-science industrial complex, from the livestock feed sector to the processed-food and beverage industries. These commodity-crop producers, many with thousands of acres of land, are the relatively elite contract peons of corporate agribusiness serfdom that has enslaved and then bankrupted hundreds of thousands of smaller, diversified family farms and crushed the spirit of rural life throughout much of the industrial world.

In a letter to his sister Wilhelmina in 1888, Vincent van

Gogh wrote, "In every man who is healthy and natural there is a germinating force as in a grain of wheat . . . what the germinating force is in the grain of wheat, love is in us." If van Gogh were alive today, I wonder how he would paint the vast fields of genetically engineered commodity crops, which look like biological deserts without a village or a field worker in sight.

Companies like Monsanto are setting a trend with "value-added" genetically engineered corn and other commodity-crop seeds.[1] Approximately 75 percent of the U.S. corn crop is fed to livestock; value-added corn, engineered to contain more oleic acid, lysine, and methionine, is supposed to make the meat healthier and tastier for consumers.

A collaborative research project between the USDA and Monsanto is making pigs grow faster and leaner by giving them frequent injections of insulin-like growth-factor and porcine growth hormone.[2] Estimates of how much financial support the U.S. government provides to this kind of production and product-oriented research for agribusiness—coupled with subsidies to producers—are around $68.7 billion per year. This translates into an extra food bill of $259 for every American consumer, according to Dr. Norman Myers.[3] This public money is used to benefit the agribusiness life-science industry by paying producers to purchase its costly, often harmful, and unnecessary products. One may wonder how this kind of agriculture can ever help feed the poor and hungry as it claims to do.

In 1998 Monsanto bought a small company that had, with U.S. government support, genetically engineered some seeds with a "Terminator gene," which creates plants whose seeds are sterile; so farmers cannot save their best seeds to use for the next season. Terminator-gene technology further protects Monsanto's interests, which are evidently not sufficiently protected under international patent and contracts with farmers that prohibit them from saving seeds to sow the next season. But it puts farmers at risk because the best seeds are those that are adapted to local conditions. The sharing and exchange of local seed varieties has been part of the sound science and culture of agriculture for millennia.

Monsanto now has a competitor for its terminator-gene technology. According to the eco-agriculture magazine *Acres USA* (November 1998, p. 2):

> The Verminator, a new chemically activated seed killer in Europe, kills seeds—in one of the invention's claims—by switching on rodent fat genes that have been bioengineered into crops. Zeneca BioSciences in the United Kingdom, the life industry spin-off of the old ICI (Imperial Chemical Industries), says it will apply for patents in 58 countries for its invention that renders it impossible for farmers to save "protected" seed from growing season to growing season. The technology, which activates a "killer" gene (or prevents the expression of genes crucial to normal plant development) weighs in whenever a chemical "trigger" is applied to seed at a desired point during plant maturation. For example, genetically engineered seed could be produced that would not germinate unless exposed to Zeneca's private chemical trigger. Or, plants growing in the field could be genetically programmed to become

stunted, not properly reproduce, or not resist disease(s) unless sprayed with Zeneca's chemical formula.

There are also legitimate concerns about the escape of genetically engineered organisms, such as trout and salmon containing the growth genes of other species that are being developed for commercial fish farming, and their potentially harmful ecological consequences. As of January 1998, the USDA and EPA had approved over four thousand releases of genetically engineered organisms into the environment for agricultural field tests. This sanguine attitude toward potential environmental risks is not rational considering the fact that newly engineered genetic traits in crops have already been shown to be transmissible to other plants, particularly weeds, and that "pests," like the diamondback moth caterpillar, quickly develop resistance to toxin-producing genetically engineered crop plants. (See chapter 7 for further discussion.)

These and other developments in agricultural biotechnology are not based on the principles of a sustainable and socially just agriculture, in spite of the promises of a safer, cheaper, and more plentiful supply of food for all. The rush to genetically engineer and patent new life forms holds great promise for investors and for those seeking a monopolistic control over the food and drug industries. But there is a clear and disturbing trend not only to misapply technology in agriculture, but also to monopolize seed stocks. This move by the multinational petrochemical-food-pharmaceutical industrial complex will mean that neither farmers nor consumers will have any control or choice over what and how crops are raised and marketed—unless genetically engineered "new foods" are so labeled in the marketplace.

Furthermore, developing new crops like soybeans and canola to produce specialty oils, and engineering bacteria to produce vanilla in fermentation vats will devastate the export markets of developing nations that produce such items as vanilla, coconut oil, cocoa butter, and palm kernel oil. Many developing nations have not attained sustainable self-sufficiency because colonial agricultural development programs have encouraged them to produce various crops for export to the industrial world, especially to feed livestock. Local people, facing hunger and poverty, then farm erodible, marginal land, destroying wildlife habitat in order to feed themselves and their own livestock. The preservation of cultural and biological diversity go hand in hand, yet industrial agriculture has been a major cause of the loss of both. When indigenous peoples lose their local knowledge, their seed stocks, and their sustainable farming practices, they become dependent upon multinational corporations as participatory consumers and contract laborers, or else they join the disenfranchised, alienated, and even starving multitudes.

Fears about the loss of genetic resources underlie the growing international acceptance of the International Biodiversity Convention, which the U.S. government still refuses to join. We need to use biotechnology wisely, to strengthen and widen the genetic base of our highly inbred varieties of cereals, fruits, and vegetables, for example. This would help them be productive under increasingly inhospitable climatic and soil conditions, which we have, in part, brought on ourselves. Seeds, like indigenous wisdom, are best conserved by being continually sown and harvested where they were developed.

The equity and social justice issues of conventional industrial agriculture are being compounded by agricultural biotechnology, which represents another costly off-farm input. Those farmers who adopt biotechnology will see their profit margins decline further as the input and market-sector profits of agribusiness increase. Farmers will likely lose all independence and become contract growers under the yoke of corporate feudalism. If this trend continues, industrial agriculture under this form of corporate socialism is likely to become increasingly dysfunctional, like the regime of state capitalism that destroyed agriculture under the former Soviet communist government.

DEVELOPMENT OF THE LIFE-SCIENCE INDUSTRY

The development of the newly self-named "life-science" industry began during the energy and oil crisis in the 1970s. The major oil companies decided to consolidate and diversify after realizing that the world's oil reserves are finite and increasingly costly to extract, and in full knowledge that many of their petrochemical products were poisoning the planet, contributing to global warming, and causing cancer and other disease in wildlife and people. Small oil and chemical companies were bought out or bankrupted by bigger ones in the industrial Jurassic Park of unbridled capitalism and transnational oligopolies.

The oligopolists and their compatriots in government and academia saw to it that a viable economic future lay in securing global control not only of natural resources, such as oil and uranium, but also of the world market, especially of the food and drug sectors. They called this monopolistic control "sustainable development." This was promoted through GATT (the General Agreement on Tariffs and Trade) under the banner of "free trade" and protected by the World Trade Organization and the TRIPS (Trade-Related Intellectual Property Rights) Agreement.[4]

In order to gain a monopolistic control of food, the evolving life-science industry of petrochemical, pharmaceutical, seed, and food companies bought up independent seed companies, eventually making only their own patented varieties of seeds available to farmers. The fit was perfect when genetically engineered seeds and farm animals were linked productively with the use of herbicides and insecticides; special feed; and production-enhancing drugs for the animals such as antibiotics, genetically engineered growth hormones, anabolic steroids, and beta-blockers.[5]

The following is from *Food Chemical News* (November 9, 1998, p.20):

> "Chemical companies see life sciences as their future. Chemistry is not the innovative science it once was," Ralph Hardy, president of the National Agricultural Biotechnology Council, said at IBC's [International Biotechnology Conference] Third International Conference on Transgenic Plants, November 2–3. "Chemicals are commodities with modest returns."
>
> In the past few years, agrichemical input companies have invested $15 billion to acquire, or form alliances with, biotechnology, seed and food companies, Hardy said. Rapid

consolidation, he noted, has led to dozens of mergers, acquisitions, alliances and licensing agreements, including deals between Monsanto and each of the following firms: Agracetus, Calgene, Ecogen, AgriPro, Asgrow, DeKalb, Delta & Pine, Holdens, Stoneville, Cargill, FirstLine and PBI; between DuPont and each of the following: Pioneer, Asgrow, Hybrinova, PTI, and CDI; between Novartis and Mycogen and Novartis and Ciba Seeds; between Dow AgroSciences and Mycogen and Dow AgroSciences and United Seeds; between AgrEvo and PGS and AgrEvo and Cargill; and between Zeneca and Advanta.

Now these multinational petrochemical and pharmaceutical companies are using biotechnology "life science" to gain monopolistic control of the world market for food and medicines. The kinds of food that are being produced, and the harmful consequences of how they are produced, have yet to be fully accounted for: aquaculture and factory animal farms and feedlots that cause tremendous pollution from animal wastes and animal suffering and create new plagues through food contamination with antibiotic-resistant bacteria; and nutrient-deficient crops from soils that have been chemically sterilized and poisoned with agritoxins. The emphasis on uniform, high-volume, commodity-crop and biomass production has put close to half a million diverse and ecologically more sound family farms out of business in the United States since the 1970s.[6]

That emphasis has also contributed to the loss of biodiversity, not only in the living soil itself, but also by obliterating natural ecosystems such as savanna grasslands, marshlands, and deciduous forests to produce various agricultural commodities. This in turn has led to serious soil degradation and to the disruption of watersheds in the United States and in many parts of the world. The farmer's dictum "We do not inherit the land, we borrow it from our children" has been ignored.

Surface groundwater pollution from farm animal wastes is a serious environmental and public health problem. Using biotechnology to correct pollution problems—so-called bioremediation—is not acceptable if no efforts are made to change the existing agricultural system and especially to raise farm animals under less intensive and more humane conditions, which can be done efficiently and profitably.

The same scenario holds true for other countries where industrial agribusiness has taken hold. The declines in food quality and safety have been of inestimable cost to the public in terms of illness, time off work, increased insurance costs, government investigations, oversight, and continued subsidies to agribusiness to boost exports.

Norman Myers estimates that the American public subsidizes this toxic food industry to the tune of almost $69 billion of their tax dollars, which their government hands out to agribusiness.[7] Many drugs, which often have harmful and costly side effects, and diagnostic services would not be needed and people would be much healthier if they consumed organically certified natural foods and beverages in moderation.

According to the Department of Agriculture's Economic Research Service, society now loses up to $7.5 billion yearly in medical costs and lost productivity from meat and poultry products from factory farms that have been contaminated with harmful bacteria. These and other health costs from various forms of cancer and immune system dysfunctions have been estimated to range from \$28.6 billion to \$61.4 billion per year by the Physicians Committee for Responsible Medicine.[8]

Public health problems from food poisoning to cancer, birth defects, infertility, and immune system dysfunctions guarantee a growing market for new drugs and diagnostics.[9] For the life-science industry to produce more healthful food in less harmful ways would, therefore, reduce profits in the pharmaceutical and diagnostics sectors. All of this, as well as the justification of much animal suffering in biomedical research laboratories, would be greatly diminished by consumer support of organic farms and whole-food marketers, as well as doctors for themselves and veterinarians for their animal companions (who often sicken and suffer from those manufactured pet foods that consist primarily of ingredients condemned for human consumption) who are not prescription pimps for the pharmaceutical industry and who put preventive medicine first.

Genetically engineered bovine growth hormone (rBGH), which dairy farmers inject into cows to boost milk production, is one product that causes an increase in a variety of animal health problems such as udder infections and foot disease, and carries possible consumer health risks (see chapter 5 for further discussion). The more economic and ecologically sound alternative of rotational grazing (by which cows are moved to a fresh pasture at intervals), as well as organic farming, is seen as a major obstacle by agribusiness in its attempts to get dairy farmers to buy this new drug. Biotechnology companies have been testing and trying to market rBGH in developing countries, which would undermine traditional sustainable livestock and forage production systems. The appropriate use of genetic-engineering biotechnology in both developed and developing countries must balance the interests of the people with those of the corporate world and not undermine good farming practices.

It is fair and reasonable to conclude that biotechnology is being applied as a band-aid remedy for the diseases and other production-related problems of both crops and factory-farmed animals whose environments are stressful and disease inducing. The same may be said of many of the applications of genetic engineering to treat various human diseases of industrial civilization.

The ethic of respect and reverence for all life needs to be applied as the DNA "code of life" is exploited. In cross section, the molecular structure of DNA is like a mandala, reflecting the sacred symmetry and mystery of life. In exploiting animals and other life forms, we should be mindful of this sacred dimension and not regard life simply as a patentable commodity or as self-replicating matter.

CHAPTER 3

Agribusiness and Biotechnology: New Crops and Food Problems

The ultimate goal of farming is not the growing of crops but the cultivation and perfection of human beings.
—Masanobu Fukuoka

The publication of a critical essay entitled "Playing God in the Garden" in the *New York Times Sunday Magazine* (October 25, 1998) by Michael Pollan is of historical note: This was the first time that a major national newspaper had given the complex issue of genetically engineered crops and new foods such coherent coverage. The skeptic in me says this article, while excellent, was too little too late in terms of informed public involvement because the gene genie is already out of the bottle, with millions of acres of genetically engineered crops now being grown and harvested around the world, from Argentina to China and Iowa to California.

The optimist in me says look at the United Kingdom and

other countries in Europe where governments, under public pressure, and with science-based and ethically considered policies are moving to ban or more strictly regulate genetically engineered crops, and to label as such all whole and processed foods that have been subjected to this technology. (See chapter 6 for more details.)

What are my main concerns, or is it just all biotechnophobia, fear of the new and unknown that the specter of genetic-engineering biotechnology and gene-altered foods can evoke? While transgenic crops that produce their own insecticides will mean that fewer harmful pesticides are sprayed on crops, we know nothing about how those insecticides will affect us or nontarget, beneficial insects, birds, and other creatures. As for engineering crops to be resistant to chemical herbicides, the use of which crops is said to reduce global warming and soil erosion because farmers don't have to plow the land to control weeds, the last thing we need is more harmful and costly chemicals on our soils.

The way desired genes are put into these plants is also of concern, since various viruses are used as vectors, to which new genes are spliced. But viruses can recombine, creating new plant and animal diseases. Foreign DNA from such viruses could be absorbed through our intestines and become incorporated in the cells of our own bodies. DNA from plant viral vectors could combine with so-called endogenous proviruses in our own cells and cause disease.

The potential disease scope is considerable, from allergic reactions to novel food ingredients and known allergens (such as peanut or brazil nut protein genes spliced into corn or beans) to neuroendocrine system disorders. Many of our own hormonal, nervous, and immune-system functions are co-evolved, plant-based, derivative biochemical processes that could be disrupted when we shift from a multimillion-year-long diet of natural plant foods to a new diet of biochemically altered, genetically engineered foods. We don't know the costs or the consequences of such foods. The seemingly simple and innocuous procedure of putting a new gene into a plant could so alter the biochemistry of that plant that people with a particular genetic makeup who ingest that plant could be harmed. The science of biotechnology is not sufficiently advanced to assure us that this is not possible.

Many genetically engineered seeds contain antibiotic-resistance markers—for example, genes that endow bacterial resistance to antibiotics like kanamycin. Some scientists are concerned that these plants, once consumed, could have harmful consequences to animals and people by transferring antibiotic resistance to other intestinal bacteria. Bacteria and other vital organisms in the living soil—of which there are four thousand to five thousand species among the ten billion organisms found on average in each gram of healthy forest soil—could also be harmed by these genes and by the new insecticides that crops produce, as well as herbicides such as Monsanto's widely used Roundup.

The advent of genetic-engineering biotechnology and its applications in crop production and food processing raise yet more questions and concerns, especially since the U.S. government has essentially deregulated this new industry ostensibly to give U.S.-based multinationals a competitive edge in the world market. The variety of genetically engineered crops now approved by the U.S. Food and Drug Administration is also increasing (see addendum at end of chapter).

FOOD QUALITY AND LABELING

Through the newly established World Trade Organization (WTO) and the Codex Alimentarius, which is drafting international agreements on food quality and safety, the prevailing values and practices of industrial agriculture, notably deficient in humane and environmental ethics, may well become the global norm. U.S. agribusiness corporations, facing international competition, understandably resist environmental and biodiversity protection legislation and regulations. It is illegal under WTO rules for the United States to protect its own farmers from imports from other countries that have inadequate or no such legislation and regulations. In the absence of international harmonization of sound environmental and ecological farming regulations and practices, food quality and safety are not likely to be achieved.

Without labeling food as to country of origin and method of production (e.g., organic, free range, or genetically engineered), U.S. consumers will have no choice in the marketplace and no opportunity to support either U.S. farmers or particular farming methods. But the U.S. government, under pressure from agribusiness, is resisting attempts by public interest organizations to uphold the consumers' right to know via appropri-

ate food labeling of all genetically engineered foods and ingredients. The soon-to-be established national organic food label
may actually set a lower standard than many U.S. organic farmers have achieved, which will set up unfair competition and
actually mislead consumers.

The claim of agribusiness that the public will not pay more
for food that is of similar nutritive value but has been produced
without harm to the land, animals, or environment is an
unfounded assumption. When the public is made more fully
aware of the harms caused by conventional agriculture, including those to human health, to the land, and to rural communities and culture and see the unnecessary but economically
rationalized suffering of farm animals, they will surely be willing to pay more, as many informed consumers are doing already.

But in reality they would probably pay *less* for food from
humane, sustainable, and organic farming and ranching systems. A full cost-benefit analysis of conventional agriculture
would show that the costs far outweigh the benefits, in which
we should include up to $60 billion annually in public health
costs related to nutrient-deficient soils and foods, harmful agrichemicals, and consuming too much animal fat, protein, and
refined, denatured, and processed foods;[1] and annual subsidies
of animal agriculture of $50 to $60 billion. Some costs and
losses—like biodiversity, wildlife habitat, and our rural communities, crafts, and cultures—cannot be given a dollar value. Nor
can the physical and emotional suffering of farm animals from
stress and disease as indicated in an annual loss of farmers'
profits at an estimated $17 billion.[2]

There is no real profit in such pointless and unethical activity. But it will continue as long as government continues to serve the interests of the life-science industry biotechnocracy; and as long as consumers do not vote with their dollars and establish community-supported agriculture to keep local producers in business, purchase produce from local marketing cooperatives, and support those nongovernment organizations that are protecting consumers' right to know by demanding proper food labeling and encouraging the adoption of more humane, sustainable, and organic farming methods.

In using agrichemicals to boost food production for profit and ostensibly to meet the demands of an ever expanding human population, agriculture has become chemically addicted. Many of the chemicals used have adversely affected not only the living organisms and elements of the soil but also the trophic processes of transmutation and energy flow at the molecular and subatomic levels. As we are killing our soils, we are doing no less to ourselves and our air, food, and water. At the molecular level of soil management and crop and livestock health and productivity, trace minerals are of special concern. Their imbalances and deficiencies are at the root of many crop, livestock, and human diseases, as they play a vital role in cellular metabolism, most enzyme processes, and all organ-system functions, especially of the immune, circulatory, nervous, and reproductive systems.

U.S. agribusiness advocates such as Dennis T. Avery would contest my concerns and criticisms of industrial agriculture, which he calls "high-yield" agriculture.[3] Its proponents, who also see biotechnology as a way to further enhance agricultural productivity and to save wildlife species and biodiversity, either deny these hidden costs or accept them because the benefits—more food (and agribusiness profits for a "hungry world")—far outweigh such costs.

The conclusion of such advocates is that because the human population is expanding and needs food, the risks and costs of intensive high-yield agriculture are justified (or insignificant). There is no alternative, such as organic farming, according to Avery, because it is so low-yield it will mean global famine if more wildlife habitat isn't taken over to make up for the deficit per acre. Thus, organic farming is seen as a major threat to conservation and biodiversity and to the human good.

The new agribusiness myth that Avery promotes is that industrial agriculture is the best way to protect the environment and biodiversity. But a recent report by the Henry A. Wallace Institute for Alternative Agriculture[4] details how, in the United States especially, chemically based, intensive crop production (especially questionable as a livestock feed source) harms both terrestrial and aquatic ecosystems; and confirms that a range of alternatives to the chemically based production model can achieve equivalent or higher yields per unit area of land with less harmful consequences.

In one of his epistles for agribusiness, Avery goes on to suggest that industrial agriculture, with its agrochemicals, agrobiotechnology, and patented hybrid seeds, will not only alleviate world hunger, but also help reduce population growth because people who have a better income and can afford more meat and other animal produce have fewer children.

This is an overly simplistic correlative inference. Smaller affluent families are, per capita, as much, if not more, of a drain on the environmental economy and energy budget as poorer families who eat little or no meat and sustain themselves via a low-input, labor-intensive agriculture.

It is education and access to family-planning programs and the development of local self-sufficiency and sustainable enterprises, especially agricultural, not agribusiness high-yield farming, that will help control human population growth and world hunger.

Agribusiness has much to contribute to help alleviate such problems as human hunger, poverty, and malnutrition and a major role to play in conservation, wildlife, and biodiversity protection; but it must be less focused on selling products, investing in, researching, and developing ever more farm inputs, since the Achilles' heel of Avery's high-yield farming is its dependence on high inputs, from chemical fertilizers to mega-farm machinery and biotechnology.

Instead, agribusiness industry should focus more on process not productivity, which is the end point of an extremely complex, biodynamic system that does not fit within the narrow paradigm of conventional agricultural economists.

By "focus on process" I mean paying attention to the economic and health benefits of maintaining a *living* soil, the primary resource—other than pure water, clean air, and normal solar radiation—of agriculture. There is much money to be made in helping restore and maintain soil, air, and water quality, as well as the quality of livestock and seedstock, without having to resort to genetic engineering. Let agribusiness find its profits in helping farmers restore agriculture communities rather than in selling more products and processes that increase farm inputs, lower farmers' profits, and increase market profits for the agribusiness petrochemical-food and feed industry complex. A science, economy, and ethics of remedial agricultural inputs that lead to healthier soils, crops, livestock, and food should be on the life-science industry's corporate agenda, and should be the primary mission of land-grant colleges of agriculture, "food science," and veterinary medicine around the world.

The same must be said for human medicine, which needs to establish a closer linkage via nutrition with remedial innovations in agriculture and in consumer eating habits. It is absurd that the pharmaceutical and medical industries should continue to profit by selling many products and treatments that would not be needed if our soils were healthy, our food was safe and nutritious, and our diets and lifestyles were tempered by the science and philosophy of biological realism and bioethics.

RESTORE THE HUMUS, RECOVER HUMANITY

The decline and demise of civilizations are almost invariably linked with the devitalization of the soil and consequent malnutrition. Today this is compounded by variously denatured, deficient, refined, processed, and adulterated foods. We human beings tend to forget that we are *humus* beings. From the earth we are born, to the earth we return, and by the earth we are sustained. *Humility, humanity,* and *humus* are words that connect and ground us in the reality of our being. But our rampant egotism, our pathological self-centeredness, separates us from the reality of our being, and out of arrogance and ignorance we neglect and abuse the earth. Caught in the delusional realm of anthropocentrism, we fail to realize that when we harm the earth, we harm ourselves. When the humus is depleted of microorganisms, when it becomes nutrient deficient and toxic with agrochemicals, so become our crops, farm animals, and the food we consume: And so become our bodies, minds, and spirits. In harming the earth, we harm ourselves physically, mentally, morally, and spiritually.

When we recover our humanity and humility, we rediscover the wisdom of living in harmony with the earth. Through the sacraments of seed and soil, and toil and food, our health and well-being and the vitality of the earth are mutually enhanced. As we enter the deep communion of a reverential symbiosis with the earth, human purpose and fulfillment gain greater meaning and significance. And we are secure in the knowledge that we are part of that which is forever being renewed, as the self is forever sustained, transfigured, transformed, and reborn. Through the inter-communion of reverential symbiosis we come to understand and respect, as the laws of nature, all the relationships and processes that maintain and sustain the life community. Obedience to these laws enables us to participate in a creative and mutually enhancing way and by so doing avoid causing harm to ourselves and other sentient beings.

As we humans come to see that our arrogance and alienation arise when we forget our origins and that most evil in the world comes from our ignorant self-centeredness, we may, with nature's help, mature into creation-centered beings. Our pathological anthropocentrism has pervaded our major religious and cultural institutions and caused great harm for millennia. The recovery of humanity and civilization lies in the anthropocosmic transformation of our consciousness, which will herald a new epoch in human evolution and in the refinement and metamorphosis of the human spirit. An auspicious beginning is to respect the living soil as a primary life giver and sustainer, and to farm and consume accordingly, with less harm and greater care, harmony, and veneration.

ADDENDUM

U.S. Food and Drug Administration Center for Food Safety and Applied Nutrition:[5] Foods Derived from New Plant Varieties Derived through Recombinant DNA Technology

FINAL CONSULTATIONS UNDER FDA'S 1992 POLICY

FDA's 1992 policy addresses foods derived from new varieties including those developed via recombinant DNA technology (commonly referred to as genetic engineering). At Calgene's request, FDA evaluated the safety and nutritional data collected by the firm and issued its decision that the FLAVR SAVR™ tomato is as safe as other commercial varieties of tomato in May 1994. Following that decision, FDA has not found it necessary to conduct comprehensive scientific reviews of foods derived from bioengineered plants based on the attributes of these products, but consistent with its 1992 policy, FDA expects developers to consult with the agency on safety and regulatory questions. FDA is requesting that firms provide a summary of their food (including animal feed) safety and nutritional assessment to the agency and discuss their results with agency scientists prior to commercial distribution. Developers have completed this process for the products listed below. Each entry represents a separate consultation, and each consultation may represent more than one line of the traits indicated. The products are grouped by the year in which their consultations were completed. The trait introduced into the variety as well as the origin and identity of the introduced gene responsible for the trait are given. Note that the listed products may have pending regulatory issues with EPA or USDA/APHIS.

1998

AGREVO, INC.

Glufosinate-tolerant soybean

Phosphinothricin acetyltransferase gene from *Streptomyces viridochromogenes*.

Glufosinate-tolerant sugar beet

Phosphinothricin acetyltransferase gene from *Streptomyces viridochromogenes*.

Insect-resistant and glufosinate-tolerant corn

The *cry9C* gene from *Bacillus thuringiensis* subsp. *tolworthi* and the bar gene from *Streptomyces hygroscopicus*.

Male sterile or fertility restorer and glufosinate-tolerant canola

The male sterile canola contains the barnase gene, and the fertility restorer canola contains the barstar gene from *Bacillus amyloliquefaciens*. Both lines have the phosphinothricin acetyltransferase gene from *Streptomyces viridochromogenes*.

CALGENE CO.	Bromoxynil-tolerant/insect-protected cotton

Nitrilase gene from *Klebsiella pneumoniae* and the *cryAI(c)* gene from *Bacillus thuringiensis* subsp. *kurstaki*.

Insect-protected tomato

The *cryIA(c)* gene from *Bacillus thuringiensis* subsp. *kurstaki*.

MONSANTO CO. Glyphosate-tolerant corn

A modified enolpyruvylshikimate-3-phosphate synthase gene from corn.

Insect- and virus-protected potato

The *cryIIIA* gene from *Bacillus thuringiensis* (Bt) sp. *tenebrionis* and the Potato Leafroll Virus replicase gene.

Insect- and virus-protected potato

The *cryIIIA* gene from *Bacillus thuringiensis* (Bt) sp. *tenebrionis* and the Potato Virus Y coat protein gene.

MONSANTO CO./ NOVARTIS

Glyphosate-tolerant sugar beet

The enolpyruvylshikimate-3-phosphate synthase gene from *Agrobacterium* sp. strain CP4, and a truncated glyphosate oxidoreductase gene from *Ochrobactrum anthropi*.

UNIVERSITY OF SASKATCHEWAN

Sulfonylurea-tolerant flax

Acetolactate synthase gene from *Arabidopsis*.

1997

AGREVO, INC.

Glufosinate-tolerant canola

Phosphinothricin acetyltransferase gene from *Streptomyces viridochromogenes*.

BEJO ZADEN BV

Male sterile radicchio rosso

The barnase gene from *Bacillus amyloliquefaciens*.

DEKALB GENETICS CORP.

Insect-protected corn

The *cryIA(c)* gene from *Bacillus thuringiensis* (Bt).

DUPONT

High-oleic-acid soybean

Sense suppression of the GmFad2-1 gene, which encodes a delta-12 desaturase enzyme.

SEMINIS VEGETABLE SEEDS

Virus-resistant squash

Coat protein genes of cucumber mosaic virus, zucchini yellow mosaic virus, and watermelon mosaic virus 2.

UNIVERSITY OF HAWAII & CORNELL UNIVERSITY

Virus-resistant papaya

Coat protein gene of the papaya ringspot virus.

1996

AGRITOPE INC.   Modified fruit-ripening tomato

S-adenosylmethionine hydrolase gene from *E. coli* bacteriophage T3.

DEKALB
GENETICS CORP.   Glufosinate-tolerant corn

Phosphinothricin acetyltransferase gene from *Streptomyces hygroscopicus.*

DUPONT   Sulfonylurea-tolerant cotton

Acetolactate synthase gene from tobacco, *Nicotiana tabacum* cv. *Xanthi.*

MONSANTO CO.   Insect-protected potato

The *cryIIIA* gene from *Bacillus thuringiensis.*

Insect-protected corn

The *cryIA(b)* gene from *Bacillus thuringiensis* subsp. *kurstaki.*

Glyphosate-tolerant/insect-protected corn

The enolpyruvylshikimate-3-phosphate synthase gene from *Agrobacterium* sp. strain CP4 and the glyphosate oxidoreductase gene from *Ochrobactrum anthropi* in the glyphosate tolerant lines. The *CryIA(b)* gene from *Bacillus thuringiensis* subsp. *kurstaki* in lines that are also insect protected.

NORTHRUP KING   Insect-protected corn

The *cryIA(b)* gene from *Bacillus thuringiensis* (Bt) subsp. *kurstaki*.

PLANT GENETIC SYSTEMS

Male sterile/fertility-restorer oilseed rape

The male sterile oilseed rape contains the *barnase* gene from *Bacillus amyloliquefaciens*; the fertility-restorer lines express the *barstar* gene from *Bacillus amyloliquefaciens*.

Male sterile corn

The *barnase* gene from *Bacillus loliquefaciens*.

1995

AGREVO INC.

Glufosinate-tolerant canola

Phosphinothricin acetyltransferase gene from *Streptomyces viridochromogenes*.

Glufosinate-tolerant corn

Phosphinothricin acetyltransferase gene from *Streptomyces viridochromogenes*.

CALGENE INC.

Laurate canola

The 12:0 acyl carrier protein thioesterase gene from California bay, *Umbellularia californica*.

CIBA-GEIGY CORP.

Insect-protected corn

The *cry1A(b)* gene from *Bacillus thuringiensis* subsp. *kurstaki*.

MONSANTO CO.

Glyphosate-tolerant cotton

Enolpyruvylshikimate-3-phosphate synthase gene from *Agrobacterium* sp. strain CP4.

Glyphosate-tolerant canola

Enolpyruvylshikimate-3-phosphate synthase gene from *Agrobacterium* sp. strain CP4.

Insect-protected cotton

The *cryIA(c)* from *Bacillus thuringiensis* (Bt) subsp. *kurstaki.*

1994

ASGROW
SEED CO.

Virus-resistant squash

Coat protein genes of watermelon mosaic virus 2 and zucchini yellow mosaic virus.

CALGENE INC.

Flavr Savr™ tomato

Antisense polygalacturonase gene from tomato.

Bromoxynil-tolerant cotton

A nitrilase gene isolated from *Klebsiella ozaenae.*

DNA PLANT
TECHNOLOGY

Improved-ripening tomato

A fragment of the aminocyclopropane carboxylic acid synthase gene from tomato.

MONSANTO CO.

Glyphosate-tolerant soybean

Enolpyruvylshikimate-3-phosphate synthase gene from *Agrobacterium* sp. strain CP4.

Improved-ripening tomato

Aminocyclopropane carboxylic acid deaminase gene from *Pseudomonas chloraphis* strain 6G5.

Insect-protected potato

The *cryIIIA* gene from *Bacillus thuringiensis* (Bt) sp. *tenebrionis.*

ZENECA PLANT SCIENCE

Delayed-softening tomato

A fragment of the polygalacturonase gene from tomato.

CHAPTER 4

Does Genetic Engineering Have a Place in Organic Agriculture?

There are alternatives to biotechnology
for feeding the world and achieving a truly
sustainable agriculture, which are worthy
goals, but the hype of biotechnology is
obscuring the path.

—Dr. Margaret Mellon

A government ploy by the U.S. Department of Agriculture to undermine the proposed National Organic Standards, which strictly prohibit the inclusion of genetically engineered organisms, was exposed by *Mother Jones* magazine on the Internet on May 13, 1998; internal USDA documents had been leaked to this public-interest magazine. According to the *Mother Jones* exposé:

> A May 1, 1997, USDA memo, written eight months before the
> department released its proposed standards, demonstrates the
> USDA's intent to ignore standards recommended by the
> National Organic Standards Board (NOSB)—a panel of

organic industry experts created in 1990 as part of the Organic
Foods Production Act. . . . The memo sheds light on the
agency's concerns about not including genetically modified
organisms (GMOs) in the proposed standards: "Few if any
existing [organic] standards permit GMOs, and their inclusion
could affect the export of US grown organic products," reads
the memo. "However, the Animal and Plant Health Inspection
Service and the Foreign Agricultural Service [USDA divisions]
are concerned that our trading partner will point to a USDA
organic standard that excludes GMOs as evidence of the
department's concern about the safety of bioengineered commodities."

The USDA clearly finds itself in a predicament: The
United States, with the support of the Clinton administration, has invested tremendous resources in biotechnology
and has become the world's evangelist for genetic engineering. In 1994 alone, the federal biotechnology research budget
exceeded $4 billion. It's estimated that over 50 million US
acres were planted with transgenic or genetically altered
crops in 1998—up from 6 million acres in 1996.

"What [the USDA wants] to do is make sure that everything possible can be done to enhance the prospects for
genetic engineering," says Margaret Mellon of the Union of
Concerned Scientists. "Biotech is facing an uphill battle for
acceptance and would . . . like to be able to say, Well, the
organic community accepted it, why don't the rest of you?"

But USDA senior marketing specialist Michael Hankin
argues that it was appropriate to include bioengineering in

the proposed standards because the agency wanted feedback from the public on its inclusion. "The department supports the [organic] industry and is responsive to the wants and needs of the consumers," he says.

According to Hankin, the push for the inclusion of genetic engineering was not only internal but also came from none other than the Clinton White House, represented by the Office of the US Trade Representative and the Office of Science and Technology Policy. US Trade Representative Charlene Barshefsky has been pressing Europe to accept bioengineered food products. And Science and Technology's presidential advisory committee includes biotech giant Monsanto's senior vice president for public policy, Virginia Weldon.

ORGANIC AGRICULTURE AND BIOETHICS

According to Ken Ausubel, CEO of the seed-savers' networking organization Seeds of Change, agribusiness today uses only 20 percent of plant varieties for 90 percent of our food, although there are an estimated 80,000 food plants. He correctly concludes that "the biggest single trigger of extinction is the introduction of hybrid seeds by the transnational corporations." He is concerned that:

> The common heritage of the gene pool is rapidly becoming privatized and concentrated in the few hands of the most powerful interests that control the world food supply, as well as fiber, medicine, and horticulture. In the last 20 years, more than a thousand independent seed houses have been acquired by major chemical and pharmaceutical companies such as

Monsanto, Dow, UK's International Chemical Co., and Royal Dutch Shell.

Such monopolism will mean that organic farmers may be forced out of business as their seed suppliers are bought up and all seeds are genetically engineered and patented, unless seed-saving cooperatives are firmly established and organic farmers' markets and food processors are supported by consumers who know and care.

A comprehensive and objective determination of whether genetic-engineering biotechnology has any place in organic farming cannot be made without a full understanding of the philosophy and practice of organic farming systems. Several bioethical criteria can be identified as the basic principles of organic agriculture, of which livestock and poultry can play an integral and integrating role in appropriate biogeographic regions.

The practice and philosophy of organic agriculture is commendable because it satisfies some key bioethical principles. First is *ahimsa*, avoidance of harm or injury: Organic farming seeks to minimize harm to agricultural and natural ecosystems of wildlife, soil microorganisms, beneficial plants, insects, birds, etc. Second is *biodiversity*: Organic farming protects and actually enhances biodiversity of both domestic and nondomestic animals and plants. And third is *transgenerational equity*: Environmental quality and the productivity of the land are secured and enhanced for future generations. These and other bioethical principles and criteria of organic sustainable agriculture, notably food quality and safety, full-cost accounting (or fair-market price), and humane husbandry are all interdependent and complementary.

The vernacular definition of *organic*, as it relates to food production, means "relating to, produced with, or based on the use of organics as fertilizers without employment of chemically formulated fertilizers or pesticides." *Organic* is defined as "a fertilizer consisting only of matter or products of plant and animal origin" (*Webster's Third International Dictionary*, 1976).

Organic farming, while in part based on the use of organic, as distinct from synthetic, chemicals to maintain soil quality and productivity, involves a number of integrated practices and components "constituting a whole whose parts are mutually dependent or intrinsically related" (*Webster's* op. cit.). This holistic and ecologistic nature of organic farming, of which there are several schools or systems and philosophies, is evident in both theory and practice. But as theories are refined and practices change, so the meanings of words inevitably change and the concept of organic farming continues to evolve.

A long-standing principle of organic agriculture is derived from ecology—namely, balance. Specifically, it is the maintenance of balance between inputs, productivity, and the agri-ecological resource base. This principle is loosely termed *sustainability* but bears the following caveat: All basic inputs (preferably from local resources) are natural/organic, rather than synthetic chemicals and genetically engineered products. All outputs and products are bioregionally appropriate, in terms of maximizing productivity and profitability without depleting the resource base beyond its capacity to regenerate naturally or be restored by appropriate human intervention. This may include the emergency use of specifically approved synthetics. Depending on the bioregion, farm animals play an integral role in helping maintain balance and optimal biodiversity.

In sum, organic farming means the use of natural and appropriate synthetic products, by-products, and processes and entails the design and adoption of cost-efficient crop, livestock, and poultry production practices that are analogous models of natural ecosystems in terms of energy flow, conservation, and sustainable productivity. It precludes the adoption of products, processes, and other inputs such as chemical pesticides, fertilizers, antibiotic feed additives, and hormone implants.

With these basic philosophical and scientific criteria concerning the structure and function of organic farming systems in mind, we will now look at the relevance and acceptability of biogenetic engineering to these systems.

TRANSGENICS AND OTHER DEVELOPMENTS IN AGRICULTURAL BIOTECHNOLOGY

Recent advances in genetic-engineering biotechnology are being developed for commercial application in animal agriculture. There are three basic approaches to enhance animal health and productivity using this new technology.

In the first, gene-spliced bacteria have been engineered to manufacture new-generation animal vaccines, and also pharmaceuticals, such as synthetic growth hormone to boost growth rates and milk yield. These latter products are claimed by manufacturers to be analogs of natural compounds already present in the animals' bodies. But the safety and efficacy of these products of biotech "pharming" await verification. They are analogous (as distinct from homologous) products—that is, not entirely natural. Their use in farm animals to artificially enhance immunity, disease resistance, growth rate, muscle mass, milk yield, etc., should be questioned in terms of improving overall animal health and well-being, since they will be utilized primarily in intensive-confinement animal-production systems. Long-term social and economic consequences to the structure and future of agriculture are also considerable.

The use of all genetically engineered production-enhancing products, such as growth hormones, in organic animal agriculture (with the possible exception of approved new-generation genetically engineered vaccines) should be prohibited on the grounds that they are non-natural, analog products.

In the second approach, gene-spliced microorganisms are being developed and soon will be marketed for feeding to pigs and poultry, and for injection into the rumens (stomachs) of cattle to help improve digestibility of feed (including such non-natural ingredients as sawdust and newspaper pulp) and to reduce nitrate levels in manure. From an organic perspective, this is wholly unacceptable, no matter what efficiencies and cost savings might be claimed. To so alter the internal physiology of farm animals by bacterial manipulation is the antithesis of organic animal agriculture.

The third approach is to develop gene-spliced farm animals, but their commercial future is at least five to ten years away. By inserting the genes of other species, or extra genes of their own kind, into developing embryos, so-called "transgenic" farm animals (including fish) have been created. Some of them are able to transmit these additional genes to their offspring. With the exception of some poultry (whose genetic lineage has been permanently changed by gene-splicing a segment of the fowl leukemia virus to convey immunity), these transgenic farm animals have been created either to be more productive, rather than disease- or stress-resistant, or to produce pharmaceutical products in their milk.

Other developments in biotechnology include embryo-transfer, cloning and DNA mapping, which have been criticized as leading to a potential loss of genetic diversity in the already threatened farm-animal gene pool, and to the selection of varieties of livestock and poultry that are suited only for intensive production systems.

Gene mapping to identify desirable and undesirable genetic traits in animals and plants is a costly and time-consuming process. Its promised benefits will be limited, however, by the reductionism of genetic determinism. In other words, the belief that gene mapping and identifying genetic markers will enable us to improve the health, productivity, and disease resistance of animals and plants is a science-based concept that may be true in theory but not in practice. Many traits that we judge good or bad involve a complex interplay of many genes, some of which are expressed only under certain environmental circumstances, or at a particular time during the organism's development or life cycle. What we judge as good traits from the narrow measure of productivity (such as egg or milk production, rate of growth, or ratio of fat to muscle and muscle to bone) may not be so good from the measure of stress and disease resistance. Traits believed to be good may not be good in different farming systems, climates, and biogeographic regions. Furthermore, the consequences of genetic screening and sequencing the genomes of domesticated animals and plants could be extremely harmful in that the resultant genetic uniformity of commercial varieties will increase the likelihood of serious disease wipe-outs as well as the development of new diseases.

This production-focused selection for commercially desirable traits using new technology may not only have undesirable biological and ecological consequences; it is also likely to have undesirable social and economic consequences, as when "improved" varieties of crops, livestock, and poultry are patented and contract growers use them, resulting in the competitive extinction of other varieties and farming systems.

Processor, retailer, and consumer demand for uniformity of produce, from apples to pork, has stimulated the loss of genetic diversity. The resultant plant and animal "genetic monocultures" are more susceptible to diseases, thus stimulating and justifying the use of agrichemicals, veterinary drugs, and using biotechnology to try to make crops and livestock more resistant to stress and disease. As Vandana Shiva concludes in *Monocultures of the Mind: Perspectives on Biodiversity and Biotechnology*, the north's approach to scientific understanding has led to industrial monoculture farming, which is foisted on the south, and which may well result in a sterile planet.[1]

BIOETHICAL DETERMINANTS

From the bioethical perspective of what I call natural philosophy, the creation of transgenic animals, plants, and other life
forms is unacceptable because such action violates the sanctity
of life and may be regarded as an act of violence. (The term
transgenic includes gene transfers that are inter-species, inter-
genera, and inter-phyla.) To change the intrinsic or inherent
nature of distinct and unique species for purely human ends
(often making them increasingly human-dependent in the
process) is unethical to those who embrace the philosophy of
reverential respect for all life. None of the ends that transgenic
life forms serve are essential or basic to our survival, but instead
serve primarily pecuniary interests. The creation of life forms
purportedly better designed to serve human ends, be it through
traditional breeding methods or new bioengineering techniques, must be opposed if those ends cause life forms to suffer,
or harm natural ecosystems or put them at risk. Because of the
lesser risk of suffering, conventional breeding techniques,
refined by genetic screening or DNA sequencing and utilizing
the untapped genetic resources of rare livestock breeds and
plant varieties through conventional cross-breeding, should
take preference over gene splicing.

The utilitarian argument that genetic engineering and
other biotechnologies could make plants and animals more productive and efficient, thus requiring less land so that more can
be saved for nature and wildlife, is unconvincing. It lacks a
bioethical framework and totally ignores the potentials of

organic and other ecologically sustainable alternative farming systems. These more natural "whole" systems are antithetical to the ethos of industrial agriculture and agri-biotechnology, in which the production of biomass commodity monocrops, and of animal protein and fat from intensive-confinement factories and feedlots, is considered progressive and not pathogenic.

From an organic and holistic animal-agriculture perspective, the creation of transgenic farm animals is an unnecessary and unacceptable alternative to traditional selective breeding and other more humane, ecologically sound, and healthful husbandry practices. Creating transgenic livestock and poultry and using other biotechnologies to make them more productive, disease-resistant, or heat or cold tolerant are not acceptable substitutes for humane farming practices, and should be opposed if they encourage intensive, "factory" production methods and other inhumane farming practices.

The major components of disease prevention are the basic principles of organic animal agriculture and the "pillars" of veterinary holistic medicine, namely: *right breeding, right environment, right nutrition,* and *right understanding and attitude* (especially by the primary caretakers). Raising animals under optimal social and environmental conditions is the best way to avert the inappropriate and wholesale use of antibiotics and other drugs (including new-generation pharmaceuticals that have been genetically engineered). It is also the best way to discredit the economic rationale for selective breeding and for even creating transgenic livestock and poultry that are better suited for intensive-confinement systems.

As these developments now stand, and judging from the direction being taken by the biotech, livestock, and poultry "improvement" ideology, transgenic livestock and poultry have no place in organic agriculture and no place in any truly sustainable farming system.

BIOENGINEERED PRODUCTS ARE NOT "NATURAL"

Agribusiness biotechnologists reason that since one of the criteria for organic farming is the use of natural products (e.g., using natural rather than synthetic chemical fertilizers), then transgenic crop varieties should be eligible for organic certification since the foreign genes they contain are from other life forms and thus are natural in origin. This same line of reasoning would accept such genetically engineered products as biopesticides and bovine and porcine growth hormone as natural and thus acceptable under organic farming and food standards. This reasoning is flawed, however, because such bioengineered products and processes either do not naturally exist in conventional crops and animals that are part of an organic farming system, or occur at much lower concentrations within the normal homeostatic range of the plant or animal's natural physiology and metabolism.

There is no scientific evidence that genetically engineered crops are the answer to world hunger. But there is clear evidence that solutions include sustainable organic agriculture, with the adoption of integrated pest-management practices and the practice of growing a diversity of crops and growing different crops (crop rotations) in different seasons. But sustainable agriculture is a threat to the agribusiness multinationals, and what they offer is the antithesis. If a farmer plants a genetically engineered crop of corn or soybeans with herbicide resistance, he can't plant any other crops unless they too are resistant to the herbicide used on the new wondercorn or supersoy.

The use of genetically engineered microorganisms for deactivating pollutants and toxic wastes through so-called bioremediation and biodegradation, to improve animals' digestive efficiency (especially of cellulose and phosphate), and to reduce waste-emission problems in livestock and poultry (such as of ruminant methane and fecal nitrates) is not organic farming. These kinds of high-cost inputs and correctives, even if they are, as living processes, analogs of natural organic processes, are not naturally intrinsic or ecologically sympatric, and their short- and long-term risks are unknown.

New developments on the horizon to modify animals' metabolism and to modify their feed so as to enhance its nutritive value, as by genetically engineering the amino acid content for specific species to create "human corn," "swine corn," or "poultry corn," may soon become a reality.

Genomic research on crops raised to feed livestock is focusing even on reducing pollution. Engineering corn, for example, that contains digestible phosphate would save hog farmers having to supplement their hogs' diets with extra phosphate, and being highly digestible would mean less phosphate in pig feces and thus less pollution, according to advocates of enhanced foods for livestock. But is this the direction to take, even engineering livestock feed-crops with new-generation proteins that may enhance animals' disease resistance, when the intrinsic problems of intensive livestock production systems and overconsumption of animal fat and protein are not addressed and actually become, in their perpetuation, a source of profit for the food and drug industry? The acceptability of these developments, along with various new feed additives, is debatable for organic agriculture. The need for such feed additives betrays the inherent limitations of intensive, animal-based agriculture, and may well have adverse animal health and ecological consequences.

Likewise, the use of new drugs such as beta-adrenergic antagonists, somatotropins, and anabolic steroids to increase "protein deposition" (as animal-production scientists call potentially pathological muscular hyperplasia in livestock and poultry) is likely to result in a host of harmful consequences, as already evidenced by dairy cows injected repeatedly with synthetic bovine growth hormone. The R&D in this field of metabolic manipulation of farm animals cannot be considered ethical or relevant to the advancement of sustainable agriculture or the attainment of food security in the near or distant future.[2]

In sum, all developments in biotechnology should, from an organic farming and sustainable-agriculture perspective, be subjected to rigorous objective and scientific evaluation on the basis of the bioethical principles and criteria described on p. 185.

CHANGING THE NATURE OF CREATION

There is a distinction between using an animal's end or telos for one's own benefit and disregarding and manipulating its telos and ethos (or intrinsic nature) for *exclusively* human benefit, as in factory farming and genetic engineering. The good farmer and pastoralist knew how best to profit from the telos of plants and animals without harming either their ethos or the ecos—the ecosystem.

This is the distinction between sustainable and nonsustainable living, and between treating animals and other life forms as ends in themselves, respecting their ethos and ecological role, and using them as a means to satisfy purely human ends. The consequences of this latter utilitarian attitude and relationship are potentially harmful economically, environmentally, socially, and spiritually. The essence of right livelihood is surely to accommodate the interests and intrinsic value of other members of the biotic community and to enable those species whom we have domesticated—as well as ourselves—to live according to their natures, with ethos, telos, and ecos fully integrated and harmoniously actualized. In sum, natural living, like natural or organic farming, has its own integrity. Ironically, in the process of "denaturing" animals under the dehumanizing yolk of industrial-scale domestication, we do no less to ourselves and suffer the consequences under the guise of "civilized" necessity and progress. But there is nothing civilized or progressive in this utilitarian attitude toward life, since it ultimately reduces the value of human life and *all* life to sheer utility, which is the nihilistic telos of the life-science industry.

ADDENDUM: THE SEVEN PRINCIPLES OF HUMANE ORGANIC SUSTAINABLE AGRICULTURE

1. Humane sustainable organic agriculture (HOSA) entails the production of domestic animal protein and fiber on the

economically prudent basis of an ecologically sound animal husbandry and the wise and appropriate use of natural resources. Such husbandry aims to enhance or at least protect the natural biodiversity of indigenous wild plant and animal species and does not result in environmental degradation and pollution.

2. HOSA is socially just, respecting human rights and interests, especially those of indigenous peoples and native, peasant, and family-farm cultures and traditions, since the preservation of cultural diversity has inherent value just as does the preservation and enhancement of natural biodiversity.

3. HOSA recognizes the connections between farm worker health and safety, consumer health, and farm-animal health and well-being. It respects the right of consumers of animal protein to wholesome and healthful produce derived from animals whose basic physiological, behavioral, and social needs and requirements, which are integral to their overall health and well-being, are fully satisfied by the methods of husbandry that are practiced. The use of veterinary drugs to maintain animal health and productivity is minimized by the adoption of humane animal husbandry practices, which in turn lowers consumer health risks.

Furthermore, animals' health and overall well-being are maximized rather than sacrificed to maximize productivity. Optimal productivity is linked with maximal animal welfare, which in turn is linked with the optimal carrying capacity of the environment and availability of renewable natural resources.

4. HOSA is bioregionally appropriate, if not autonomous, linking livestock and poultry production with ecologically sound, organic crop and forage production systems and/or environmentally sound rangeland management.

5. HOSA does not engage in the import or export of any agricultural commodities, especially meat, wool, hides, and animal feedstuffs, that have been produced at the expense of natural biodiversity and nonrenewable resources, and that undermine the rights and interests of local farmers and other indigenous people who practice sustainable, ecologically sound, and socially just agriculture.

6. Philosophically, HOSA is based on the aphorism that we do not inherit the land, we borrow it from our children; it is ours only in sacred trust. This means, therefore, that HOSA entails respect and reverence for all life, its philosophy being creation- or earth-centered. It therefore embraces concern for the rights and interests of people, animals, and the environment. By so doing, it reconciles conflicting claims and concerns with the absolute right of all life to a whole and healthy environment and to equal and fair consideration.

7. HOSA provides the foundation for a community of hope and of a planetary democracy, whereby world peace, justice, and the integrity of creation may be better assured. It leads to the recovery of culture, agri-*culture* being the cultivation of the land and the production of food based on a hallowing covenant that commits us to the sacred obligation of caring for the earth by farming with less harm and eating with conscience.

CHAPTER 5

Genetically Engineered Animals and the New "Pharm" Animal Factories

A cow is nothing but
cells on the hoof.
—Dr. Thomas Wagner

TRANSGENIC ANIMALS

Animals carrying genes from other species that have been inserted during early embryonic development are referred to as *transgenic*. Transgenic organisms carry genetic information not normally present in the species. This genetic information has been deliberately amplified, spread, or disseminated in the species at a much faster rate than occurs naturally. In many instances these new genes will be passed on to offspring, so they become permanently incorporated into the germ line, or hereditary makeup of the animal.

Thousands of varieties of transgenic animals have been created, and in many instances mortality is high and suffering

considerable. The first transgenic animal to be patented was a
mouse that was bioengineered by Harvard University scientists
with funding from DuPont Chemical Company to be
extremely susceptible to cancer-causing chemicals and to succumb at an early age to breast cancer. The first transgenic farm
animals were developed with public funds by scientists of the
U.S. Department of Agriculture, who spliced human growth
genes into pigs. These pigs experienced considerable suffering,
physical deformities, and high mortalities at an early age.

The genetic engineering of animals produces a number of
questionable consequences. These include physical suffering;
psychological distress; behavioral impairment; developmental,
reproductive, and immunological disorders; metabolic and
regulatory disturbances; different nutritional and other physiological requirements; and new diseases that will be difficult to
diagnose and treat. There are also moral concerns such as the
sanctity of life and the integrity of species, and ecological concerns, as with the potentially adverse environmental impact
and harm to wild plant and animal species and communities if
deliberately or accidentally released transgenic animals, especially insects and fish, become established and multiply. This
threat to natural biodiversity is compounded by the increasing
loss of genetic diversity in domesticated plant and animal varieties as a shrinking genetic pool is utilized in conventional
agriculture. Contrary to the claims of genetic-engineering
advocates, this technology will not increase genetic diversity. It
will more likely result in unforeseen genetic anomalies and
increased susceptibility to environmental stress and diseases

because of the inherent genetic uniformity of existing commercial plant and animal breeding stocks that are now being genetically engineered.

"PHARM" ANIMAL BIOREACTORS*

The results of various genetic engineering experiments reviewed by Vernon G. Pursel, Ph.D., a USDA employee and one of the creators of the crippled transgenic "Beltsville pigs,"[1] make dubious a future in which biotechnology is used to make farm animals more productive. Pursel states, "At present, the transgenic approach for improvement of farm animals for production purposes remains only a hope for the future." As for engineering farm animals to produce new-generation health-care products, he cautions:

> Even though several human therapeutical proteins have now been successfully produced in milk and blood of transgenic animals, some difficult problems must be solved before these products are approved for use. Product safety is a large issue. These products will require the same rigorous scrutiny as the products extracted from animal tissue produced by tissue culture or synthesized by recombinant organisms. Products from transgenic animals must be purified to remove all non-human proteins that might cause allergic reactions. In addition, it is still not known whether these complex human proteins are sufficiently similar in structure and biological

---

*Bioreactor is the term used by the industry for an animal or animal's organ that produces pharmaceutical products.

activity to the natural proteins produced by the human body so that antibodies are not produced. While scientists are confident that these technical and regulatory obstacles can be overcome, few people are willing to predict how long it will take to work out these problems and complete the clinical testing that will be required to obtain approval of regulatory authorities for marketing to the public.

Scientists of the U.S. company Genzyme Transgenics Corporation reported a "biopharming" breakthrough in January 1998 with the creation of transgenic mice that produce human growth hormone from urinary bladder cells. This is seen as offering an alternative to the mammary gland as a bioreactor for producing pharmaceuticals, since milk, unlike urine, contains much protein and fat, making purification more difficult and costly. Plus with a bladder bioreactor, products could be harvested shortly after birth from both sexes. Also, milk may contain more viruses and other potential pathogens. Since plants, unlike goats and cows, have few diseases that are transmissible to humans, creating transgenic plants to produce vaccines and various pharmaceuticals would seem safer and less costly, and the plants would be easier to distribute in developing countries, as by growing transgenic bananas and spinach locally to produce rabies, hepatitis B, and other oral vaccines for humans and animals.* The creation of such transgenic plants,

---

*The first successful trials of an edible vaccine (to prevent *E. coli* travelers' diarrhea) in genetically engineered raw potatoes was announced in April 1998 by the Boyce Thompson Institute for Plant Research, Inc., an affiliate of Cornell University, in collaboration with the University of Maryland School of Medicine's Center for Vaccine Development in Baltimore.

provided they do not cause genetic pollution and are protected from nontarget species, is one area of genetic engineering I would endorse in these times of plagues and pestilence.

One ethical issue concerns the ever intensifying commercial exploitation of animals as bioreactors or "biomachines" and as a source of replacement body fluids and parts for humans, from blood and bones to livers and hearts. These valuable, patented human creations—"manimals"—of the new industrial biofarms of the next century will serve a wealthy elite. Their existence may deter some extremely wealthy people from having themselves cloned as a source of organ parts. Manimals instead will provide replacement parts as needed. But will their products be safe?

Transgenic animals do produce relatively higher concentrations of monoclonal antibodies and other health care products in their milk, such as lactoferrin, than can be achieved using cell cultures and transgenic plants. But this advantage of higher production efficiency must be weighed against the greater risks of disease (zoonosis) from animals to humans, from retroviruses[2] to prions,[3] a bovine variant of which has caused an epidemic of a human brain disease in the U.K. called Creutzfeldt-Jakob disease. A major advantage of using transgenic plants rather than "pharm" animals to produce pharmacological proteins is that the likelihood of cross-contamination or rejection by the human immune system is lower with plant proteins. Regardless, European-based Pharming B.V. is developing transgenic animals to produce the following biomedical proteins in their milk: human lactoferrin/lysozyme, human collagen, and human serum albumin. PPL Therapeutics (in

Blacksburg, Virginia, and Edinburgh, Scotland) is developing animals to produce the human protein alpha lactalbumin.

ORGAN PARTS

According to the *London Times* (October 26, 1997), while pig xenotransplants (genetically engineered pig organs designed to be put into humans) are banned in the U.K., three hospitals in the United States are connecting patients with hepatic failure to transgenic pigs' livers.*

The August 1998 issue of the British medical journal *Lancet* carries a cautionary word in its "Early Reports" section, "Expression of pig endogenous retrovirus [PERV] by primary porcine endothelial cells and infection of human cells," in which Ulrich Martin and his coauthors state that pigs pose "a serious risk of retrovirus transfer after xenotransplantation" (retroviruses are a type of virus that lives in animals' cells and can be inherited). In three breeds of pig from twelve sites in Denmark, Russia, Germany, and France, these investigators detected PERV in every sample of skin, liver, lung, and aortic endothelial cells. Co-cultivation of the aortic cells with human embryonic kidney cells "led to productive infection of the human cells and expression of PERV," they wrote.

In a review in *Nature Magazine* (vol. 391, 1998, p. 322), further concerns about the hazards of human infection with pig

---

*Controversy was sparked in May 1998 when Imutran, the U.K. company that has developed organ-donor pigs, shipped some to the Netherlands to a primate research facility where macaque monkeys would be experimented upon as recipients of pigs' organs.

viruses from infected organs was underscored by the finding that ten diabetic patients who received pancreatic islet cells from pigs had antibodies to porcine influenza virus, five to parvovirus, and five to other pig viruses.

GENETICALLY ENGINEERED POULTRY

Because chickens breed quickly and their embryos beneath the shell are easier to manipulate than mammalian embryos, it is surprising that so little transgenic research has been done on them. Engineered eggs could give new meaning to nutraceuticals (nutrients with medical benefits) and provide human pharmaceuticals such as immunoglobulins and blood proteins. Currently, modified avian leukemia and reticulo-endothelial cancer are used as vectors to transmit genes, many being human, into developing chick embryos and adult birds. According to chicken engineer Dr. F. Abel Ponce de Leon, head of the Department of Animal Science at the University of Minnesota, Minneapolis, whose work is supported by a grant from Sima Biotechnology, a subsidiary of Avian Farms:

> Chickens produce sugar moieties [types] more similar to those produced by humans than those made by cattle. There may be a large number of pharmaceuticals that are better to produce in the chicken—for instance, thrombosis drugs and blood thinners. Because there is no biological cross-reactivity, overexpression of these drugs won't harm the chicken. These are theoretical advantages. We won't know for sure until we have a finished product.[4]

Creating chicken and egg bioreactors to produce useful health care products may be preferable if Dr. Ponce de Leon is correct that transgenic mammals can be harmed by overproduction of these drugs in their bodies.

The problem of "overexpression" of transgenic traits in animals is a serious animal health and welfare concern. Excessively high levels of growth hormones in transgenic sheep developed by Australian government scientists, for example, induced a diabetic condition that led to ketosis and death; while in pigs, serious joint problems, stomach ulcers, and infertility were reported by U.S. government scientists. The diagnosis and treatment of these new kinds of disease in transgenic animals will be a difficult and costly challenge.

ENGINEERING MORE MEAT

One of the most disturbing areas of transgenic animal research is aimed at creating massively muscled farm animals that are essentially crippled monstrosities, many of which have to be delivered by Caesarian section because they are already so enormous at birth. A mutation in the myostatin gene causes double muscles in Belgian blue cattle. When this muscle-growth–regulating gene was "knocked out" in engineered mice, their muscles grew huge. Research into this technique is proceeding in the hope of making farm animals like these mice, with the rationale being that "extra helpings of tasty meat at essentially no cost could prove hard to resist," according to reporter Steven Dickman.[5]

Following another transgenic-mouse model that resulted in enlarged muscles and reduced body fat when the chicken cSKI gene was transferred (a system developed by bioengineers at USDA and patented by the U.S. Chamber of Commerce), pigs and calves were similarly engineered. The results were catastrophic for the animals—abnormal, crippling muscle growth, muscle atrophy, weakness, and degeneration.

The *London Times* of May 14, 1995, reported that Israeli scientists have bioengineered broiler chickens with 40 percent fewer feathers, which eat more because they are cold and hence grow faster and go for slaughter sooner; that Australian scientists have developed a "self-shearing" sheep that sheds its wool; and that USDA's second-generation Beltsville, Maryland, "Schwarzenegger pigs"—made to grow massive muscles following transgenic insertion of a suspected chicken cancer gene— became sickly cripples by three months of age.

We now have the technology to clone beef, chicken, and pork cells and to make nutritious biological analogs of hamburgers and pork sausages without causing animals any harm. So why don't we use this technology instead of raising billions of animals in bioconcentration camps—the factory sheds and feedlots that blight the countryside?

The animal factories of agribusiness operate under the economics of bioconcentration, coupled with centralization of production and product uniformity. Ecologists and forensic scientists recognize bioconcentration, the overpopulation of various species, as an indicator of population imbalance and disease. This can mean the demise of biodiversity and healthy ecosystems. Imbalance creates dis-ease, and this is why farm animals become sicker and suffer the more we subject them to bioconcentration in factory production systems and extensive rangeland conditions of overgrazing. The state of mind that embraces such bioconcentration as necessary and even progressive must submit to nature's wisdom, or suffer the consequences.

Turn from the pig, chicken, cow, and bullock concentration camps to the bioconcentrated and chemical-dependent fields and plantations where genetically "improved" biomass commodity crops, such as wheat, rice, corn, cotton, tomato, tobacco, and potato—our life's staples—are now being planted. These new "supercrops," because of bioconcentration and biological uniformity, will be subject to a host of new diseases, fungal blights, viral wilts, and insect invasions. The same fateful consequences of bioconcentration, coupled with an accelerating loss of biodiversity and cultural diversity, will afflict us ever more severely. Thus, the less choice we will have in the marketplace for natural, wholesome food.

Had her forests not gone to beef and lumber, Honduras would not have been so devastated by the 1998 hurricane. Nor would millions continue to sicken and even die from eating the products of bioconcentrated industrial fields and animal bioconcentration camps, which are often contaminated chemically, bacterially, and in other ways.

And it is cause for concern that China, whose rapidly industrializing human biomass is over one billion, with India close behind, should now begin to worsen the bioconcentration and biodiversity situations by adopting Western systems of livestock and crop production. China now has millions of acres of genetically engineered crops, especially tobacco, for export. More Chinese are eating chicken and hamburgers and are already experiencing the rush of Western appetites and disease. Humanity should be advised that until all living beings under our control are seen as subjects to be served rather than as objects to be exploited, we and they will never be well.

HORMONE STIMULATION OF DAIRY COWS

In 1994, the U.S. Food and Drug Administration approved farmers' use of the genetically engineered hormone rBGH, designed to increase milk production in dairy cows, in spite of widespread protest from consumer, farmer, and animal-protection organizations. The American Medical Association said the milk was safe, and consumers were denied the right to have all dairy products from injected cows so labeled.

In March 1994, the Clinton White House published an eighty-page report, *Use of Bovine Somatotropin (BST) in the United States: Its Potential Effects*, a pro-industry document that concluded, "There is no evidence that BST poses a threat to humans or animals."

Later in 1994, British scientists told news media that their attempts to publish evidence that rBGH may increase cows' susceptibility to mastitis (an udder infection that can put consumers at risk) was blocked for three years by Monsanto.[6] They showed that Monsanto's analyses, which were given to the FDA to show no harm to cows or increase in mastitis, were based on selected data that covered up what their analyses had actually revealed—more white cells (pus) in rBGH-treated cows—a 191

percent increase compared to untreated cows. The animal
health and welfare concerns that opponents of rBGH foresaw
have since become a reality, and consumer health concerns
have not been answered.

According to the Wisconsin Farmers' Union, over eight
hundred farmers using rBGH have reported health problems
with their cows, and adverse reaction reports filed with the
FDA jumped over 800 percent between September 1994 and
March 1995. Reports of side effects encountered after injecting
cows with rBGH have included death, serious mastitis out-
breaks, hoof and leg problems, and reproductive problems,
including spontaneous abortions.

A New York farmer publicly denounced rBGH after losing a
quarter of his dairy herd. A dairy farmer in Florida stopped
using rBGH after nine of his cows died. His herd has yet to
recover, and his overall production is still suffering.[7]

In December 1995, several Canadian newspapers reported
that Monsanto tried to bribe Health Canada government offi-
cials with several million dollars to approve rBGH.

Although rBGH is being used less and less by dairy farmers
because of cow health problems, milk from rBGH is still marketed without being labeled as such. An estimated one-third of
all cows in the United States were injected with this hormone
in 1997. Also, much research is being done on pigs to develop
an injectable genetically engineered porcine growth hormone
to make them leaner and grow faster.[8]

Two experienced investigative journalists, Steve Wilson and
Jane Akre, who uncovered too much about Monsanto's rBGH

product Posilac were fired from WTVT, a Fox TV station in Tampa, Florida, in June 1998. The Fox News Network allegedly caved in to Monsanto's lawyers, and significant findings from the investigation were suppressed.[9]

According to recent polls, 80 percent of U.S. consumers are concerned about the human health hazards of foods from cows injected with rBGH. Consumer and farmer support for rBGH-free products is evident not only in the United States, but also in the European community and Canada. The European Union, which includes fifteen countries, has extended its moratorium on the commercial use of rBGH through 1999. In 1997, the World Health Organization's Codex Alimentarius ruled against approving rBGH for use in dairy cows because of possible animal and human health risks.

An article on May 9, 1998, in *The Lancet* showed that the rate of breast cancer is up to seven times higher in women with a relatively small increase in blood levels of the growth hormone Insulin-like Growth Factor I (IGF-1). Elevated IGF-1 levels have also been correlated with other major cancers, particularly colon and prostate. And the January 1996 issue of the *International Journal of Health Services* reported that IGF-1 concentrations are up to ten times higher in rBGH milk. As IGF-1 can be absorbed through the intestine, scientists are concerned that drinking rBGH milk could increase the risk of cancer.

In poll after poll, 70 to 90 percent of U.S. consumers indicate that they want the milk that comes from rBGH-injected cows to be labeled as such. Yet the public continues to be denied its right to choose rBGH-free dairy products.

The FDA's approval in 1993 of Monsanto's rBGH and of the milk from cows injected with this genetically engineered hormone was apparently based on insufficient data. According to *Rachel's Environment and Health Weekly* (October 22, 1998)[10] the Consumers' Union and other U.S. consumer groups have called for a full congressional investigation. Apparently, the Canadian government obtained, under the Freedom of Information Act, critical data on rat studies that were not provided to the FDA or published in a study of the safety of rBGH, which showed that rats that were fed rBGH developed antibodies to this hormone, and male rats developed cysts on the thyroid gland and abnormal histology in the prostate gland.

The FDA apparently based its decision on a summary of safety studies submitted by the company, which did not include these serious findings. The Canadian government also released data concerning the health and welfare problems of cows injected with rBGH, including inflammation of the udder, which may have an impact on human health, and concluded that, "There is insufficient information [about increased levels of insulin growth factor in the milk] to provide a quantitative risk assessment, therefore, many potential health concerns remain unresolved."*

Since consumers have the right to know, and for a variety of humane and ethical reasons may wish to avoid certain foods, all food products derived from genetically engineered animals and crops, and food ingredients (such as genetically engineered

---

*In January 1999, the Canadian government prohibited rBGH because it could be harmful to cows.

yeast and rennet) that have entailed the use of some genetically engineered process, or product (such as milk from rBGH-injected cows) should be so labeled.

Using genetic-engineering technology (GET) to enhance the productivity and food-conversion efficiency of farm animals and fish is based on an industrial approach to food production, which is being questioned today, especially in terms of economic sustainability and ecological viability. Using GET to increase animal produce is also at odds with the mounting nutritional evidence that a reduction in the consumption of meat, eggs, and dairy products is advisable for health reasons.

BRING ON THE CLONES

In March 1997, the news media around the world splashed the story of a British scientist, Ian Wilmut, who had succeeded in making the first clone of a mammal from an adult cell. Wilmut used a cell taken from the udder tissue of a sheep fused into an empty ovum or egg, which was then implanted into the womb of a surrogate ewe. This replica, called Dolly, along with her creator, became an instant international celebrity.* Some likened Wilmut's achievement to those of Galileo, Einstein, and Copernicus. But cloning is no worldview-changing discovery. Other nonmammalian creatures such as frogs have been cloned

---

*The scientific paper giving details of the experiments that produced Dolly was published later in Science (December 19, 1997). Of the fourteen cloned fetuses alive at day sixty of pregnancy, only five developed into lambs that survived for more than two weeks after birth—a failure rate of 64 percent (the usual mortality rate of lambs after normal breeding is 8 percent). Some lambs were stillborn; one had a heart defect, and was euthanized at fourteen days old.

by biologists in the past, and for the vast majority of life forms on earth, it is the asexual way of achieving species survival and multiplication. This new technology, however, was so newsworthy because it is disturbing to the public, bringing us to the threshold of human cloning—which has already been done with human embryos by scientists in South Korea (and probably in other laboratories in other countries), but embryos were only developed in vitro and not allowed to grow more than a few cells. Critics linked this new technique with Mary Shelley's *Frankenstein* story, reporters asked if Dolly has a soul, and three out of four Americans in a Time/CNN survey said they believed such research is "against the will of God." Significantly, perhaps, when asked by a reporter at a 1997 Senate subcommittee hearing on cloning if he held any particular religious belief, Dr. Wilmut said that he categorized himself as an agnostic with no particular religious belief or theology.

What is the point in cloning animals, and where might it all lead? One immediate fear is that humans will be cloned once the technique is perfected. Only 29 of some 277 sheep embryos cloned by Wilmut developed normally. Many died before birth, had defective kidneys, or were abnormally large—not the carbon-copy replicas the biotechnologist had anticipated. Concern about the welfare of such animals, as with transgenic animals, is neither effectively regulated nor addressed in most research reports.

The U.K.'s Roslin Institute, where Dolly was created, quickly moved to patent its proven cloning technique. Rural Advancement Foundation International comments as follows on the Roslin Institute's world patent on the cloning of all animal species, including humans (WO 9707668, WO 9707669):

> The UK's Roslin Institute is so sure it has an economic winner
> it is claiming its cloning patents in even the weakest of
> economies—North Korea and Liberia, for instance. The
> patents are licensed to PPL Therapeutics, a company which
> has agreements with major drug multinationals like Novo
> Nordisk, Boehringer Ingleheim, and American Home Prod-
> ucts. More licenses may be granted. Unlike many bioengineer-
> ing patents, which are specified for "nonhumans," Roslin says
> its cloning patents cover all animals, including humans.

Soon after the news splash about Dolly, other sheep, cows, pigs,
and monkeys were reported to be pregnant with clones created
by methods similar to those used to make Dolly the sheep
(*Washington Post*, June 28, 1997). Several researchers reported
high embryo mortality (miscarriages) and abnormally large size.
Veterinarian Mark Westhusin at Texas A&M University stated,
"We've had some [calves] that were just monstrous, up to 180
pounds," compared with the normal 80 pounds. Oregon Primate Research Center reported creating two rhesus monkey
clones and had several monkeys pregnant with clones.* Scientists in many fields, including AIDS research, have already put
in requests for such monkeys.

The specter of cloned animal suffering is, however, very real.
Dolly-like clones at the Roslin Institute that died soon after
birth were larger than normal, putting their mothers at risk, and
had congenital abnormalities in their kidneys and cardiovascular

---

*These two monkeys were not true clones but were made by splitting embryos. No
true clones of monkeys have yet been created (*Washington Post*, May 10, 1999).

systems—problems not reported in the scientists' first paper in *Nature* magazine.[11] Miscarriages may be due to improper development of the placenta. In recent studies of Dolly, Dr. Wilmut has found subtle changes in chromosome structure usually found in cells of older animals—evidence of a possible "molecular memory," as the cell that was used to create Dolly came from a six-year-old sheep.

Likewise, it was some time after the creation of giant "supermice" by R. D. Palmiter and coworkers that many health problems were reported, notably chronic kidney and liver dysfunction; tumor development; damage to female reproductive organs; structural changes in the heart, spleen, and salivary glands; plus shorter life spans and high infant and juvenile mortality.[12]

In a press release from Tass News Agency, Moscow (July 29, 1997), over one hundred new types of animals have been cloned in Russia, opening "boundless opportunities foremost in intensive industrial cattle-breeding," as well as pig organs for humans and new varieties of sheep that have biopharmaceuticals in their milk to treat gastro-intestinal diseases and more quickly ferment sheep milk into cheese. China has reported cloning of pigs, rabbits, and cattle, and researchers in Australia and Denmark are developing cow clones.

In September 1997, PPL Therapeutics of Edinburgh, a corporate offshoot of the creators of Dolly, produced a sheep called Polly. She was a clone like Dolly but was also the first transgenic animal to be cloned; she bore a human gene that would cause her to produce alpha-1-antitrypsin, a human blood protein used to treat cystic fibrosis, in her milk. Of two other live transgenic clones produced at the same time as Polly, from a total of sixty-two embryos implanted into surrogate mothers, one died soon after birth. Also, some births had to be induced or performed by Caesarian section because, for unknown reasons, natural labor failed to occur, according to a *Washington Post* report (December 19, 1997).

PPL's American Division in Blacksburg, Virginia, soon after announced its success in creating three transgenic rabbits, whose milk contains calcitonin (from an inserted salmon gene), a bone-building substance that can be used to treat osteoporosis. The company plans to try this procedure in farm animals. Around this same time, ABS Global, Inc., of DeForest, Wisconsin, unveiled a healthy six-month-old calf named Gene developed with its own patented cloning technology.

The *Veterinary Record* of October 18, 1997, reported that scientists at the Roslin Institute are now particularly interested in creating in sheep the same range of mutations that give rise to cystic fibrosis in humans. Asked whether he felt it was morally correct to introduce genetic defects into an animal, Dr. Wilmut (Dolly's creator) said that, provided the animals receive the same degree of care as a person with the disease and there was otherwise no realistic prospect of bringing forward treatment, he would be "quite comfortable" to be involved in such work.

On a BBC TV *Horizon* documentary entitled "Hello, Dolly" (October 23, 1997), the founders of Granada Genetics in Texas, who were the first to market cloned cattle, admitted that their venture had failed because many bovine clones were abnormally large and had to be delivered by Caesarian section, had enlarged hearts, and developed diabetes.

Scientists at the University of Hawaii announced in 1998 the birth of a third generation of mice cloned from adult cells, demonstrating the most successful technique to date. But many gene engineers insist the technology still has far to go before it is commercially viable, with only a 1 to 2 percent success rate of eggs injected with an adult animal's cell. An article in *Nature Biotechnology* (September 1998, p. 809) included the following:

> "While we know how to do nuclear transfer with adult somatic cells, we don't know how to do it efficiently," said Ian Wilmut at the recent Second Annual Congress on Mammalian Cloning in Washington, DC in late June. "There are losses at every stage of development of the embryos, as well as serious congenital malformations in many animals."
>
> This was echoed by a number of speakers at the meeting, who noted that "a great number" of animals born from nuclear transfer have been born missing organs such as kidneys and hearts, and that many have also been stillborn or born greatly oversized. "Through our current methods of nuclear transfer, we don't really know what kind of embryos we are creating—normal or not," said Tanja Dominko, staff scientist in Gerald Schatten's laboratory at the Oregon Regional Primate Research Center (Beaverton), and who used to work on bovine cloning at the University of Wisconsin.

Dominko and Dr. Neal L. First of the University of Wisconsin have emptied out cow ova and reportedly succeeded in getting cells from the ears of mature pigs, sheep, and monkeys to become embryonic inside these ova. None implanted, however, when put into the uterus of a surrogate animal mother.

Dr. First had hoped this technique could provide a way to clone endangered species, but now other biotechnologists are developing this technique of using emptied-out cows' eggs to contain fused human embryo cells, called stem cells. The aim is to clone various human organs, as well as brain and heart muscle cells, for transplant into humans. Patients' own cells could also be cloned to reduce chances of immune rejection, according to Jose Cibelli of Advanced Cell Technology in Worcester, Massachusetts (*New Scientist,* July 11, 1998, pp. 4–5). Viable human embryos would not be produced in this process, so one ethical hurdle would be avoided. But the use of cows as surrogates in the production of spare parts for humans raises the specter of an increasingly parasitic human relationship with and dependence on animals that is aesthetically disturbing, if not biologically regressive.

A new way possibly to create spare organ parts has been described by Dr. Jonathan Slack of Bristol University. Slack has controlled frog egg genes to create headless tadpoles. He says his research could lead to the production of cultured human organs such as a heart or pancreas in reprogrammed, cloned human eggs. Conceivably headless humans, like his headless tadpoles, could be cloned, but with perfection of this technology, specific body parts could be created from cloned human cells.[13]

A new technique using calf embryonic stem cells to create clones was announced in June 1998 by researchers at the University of Massachusetts, who produced twelve calf clones. Their company, Advanced Cell Technology, Inc., has contracted with Genzyme Transgenics to develop genetically engineered cows that produce in their milk a human protein that is used to treat blood loss, using the cloning technique to rapidly build up a uniform herd of transgenic animals. France has also joined the cloning craze, showing a video of "Margueritte," a calf cloned from a muscle cell of a calf fetus, at the Paris Agricultural Show in March 1998.*

In the fall of 1998 the biggest set of clones so far from one animal at one time were produced—ten calves from cultured cells of one cow were born in New Zealand at the government's Agricultural Research Center in Ruakura. Around the same time, Snow Brand Milk Products (SBMP, Hokkaido), one of Japan's largest dairy companies, succeeded in impregnating cows with embryos cloned from mammary gland cells extracted from cow's milk. If successfully born, the calves would be the first animals cloned from somatic cells collected from milk.

One real problem with creating animal clones is their genetic uniformity, which is likely to decrease their survivability by increasing their vulnerability to infectious diseases. Another problem is the suffering of those animals afflicted by genetic and developmental defects caused accidentally because the technology is not risk free or caused deliberately for biomedical research into human genetic and developmental diseases.

---

*Fresh fears about cloning humans were raised when this calf died from anemia at seven weeks because her immune system never developed properly (Renard, J. P., et al. *The Lancet* 353 [1999]: 1489–91.)

If cloning biotechnology is ever perfected, what will it mean for humans and other animals? The technique developed by Wilmut has been patented, so venture capitalists have high hopes that cloning will be a boost to the organ transplant industry and to pharmaceutical "pharming" of health care products (see table on p.117). Pigs, sheep, cattle, and goats have already been genetically engineered, respectively, to serve as organ donors for people, to produce more humanized milk, and to produce milk containing valuable bio-pharmaceuticals. The numbers of these animals might now be rapidly increased using cloning biotechnology.

GOVERNMENT HEARINGS

National Institutes of Health Director Dr. Harold Varmus warned a congressional subcommittee in 1997 that had called a hearing on this controversial and wrongly overblown cloning issue. (I say "wrongly overblown" because at both House and Senate hearings and in most media coverage, there was no mention about how this may harm or benefit animals.) Varmus, pointing out correctly that much of the discussion is way ahead of where the science is, said that it could take just one infertile couple to opt for cloning, if that was the only way they might have a child, to make human cloning a reality. At this time (around the Ides of March), President Clinton proclaimed that "each human life is unique, born of a miracle that reaches beyond laboratory science," and immediately banned the use of federal funds for human cloning. Congressman Vernon Ehlers (R-Michigan) introduced two anti-cloning bills. But Senator Tom Harkin (D-Iowa) told the Senate subcommittee hearings, convened by former heart-transplant surgeon Senator Bill Frist (R-Tennessee), that he vehemently opposed the president's cloning moratorium, proclaiming, "Human cloning will take place in my lifetime, and I don't fear it. I welcome it, its untold benefits."

But Dr. Ian Wilmut, a slight, balding, bespectacled Englishman in his mid-forties, challenged the senator. This brave new animal bioengineer had come from the Roslin Institute, close by the wild moors and glens of Scotland, where a few natural sheep and shepherds try to hold on to a probably more sustainable way of life than this new technology could ever offer. He gave a rebuttal to Senator Harkin, stating that he could see no reason to ever engage in the cloning of human beings and that an international moratorium should be put in place. Dr. Wilmut is a product of the late-twentieth-century industrial revolution that is fusing commerce ever more destructively and intimately with the earth's creation. This dominant culture is exploiting biodiversity, rushing for plant and other life patents, patenting transgenic animals, and even patenting human cells and rare genetic types immune to certain diseases.

I was in India, helping my wife at the Mavanhalla Animal Sanctuary, when the news of Dolly the clone made international headlines. Following this, there were news reports of an Indian transplant surgeon who claimed to have successfully grafted the heart of a pig into a patient (probably a sick "human guinea pig" from a shanty slum) using "an antigen-suppressing agent" he claimed to have developed. However, the patient died of an infection a few days later. This doctor claimed the patient was sick with the infection prior to the transplant (*Times of India*, March 3, 1997). Because of the risk of transplant recipients becoming walking time bombs for new virus diseases that they might develop from contaminated animal organs, a moratorium on xenografting has been put in place in Europe. But there's nothing to stop a "mad" scientist from subjecting humans to biotechnological "enhancement" and self-replication through cloning.

TRANSGENIC BIOPHARMACEUTICALS

(Proteins successfully expressed in transgenic animals by Genzyme Transgenics Corporation)*

Proteins	Disease Targets and Other Purposes
Antithrombin III	Blood clots
Alpha-1 proteinase inhibitor	Hereditary deficiency, cystic fibrosis
Angiogenin	Cancer
Beta interferon	Cancer, multiple sclerosis
Glucocerebrosidase	Gaucher's disease
Human serum albumin	Hypoproteinemia
Insulin	Diabetes
Long-acting tissue plasminogen activator	Heart attacks, wound healing

*From the statement given at the Senate Subcommittee on Public Health Safety Hearings on Cloning, Implications and Risks, March 12, 1997, by Dr. James A. Geraghty, president and CEO of Genzyme Transgenics Corporation.

Myelin basic protein	Multiple sclerosis
Prolactin	AIDS, food supplement
Urokinase	Blood clots
GAD	Diabetes
Mab-fusion protein	Cancer, drug delivery
Mab-tumor marker	Cancer, drug delivery

ENGINEERING ANIMAL DISEASE "MODELS"

The arguments in favor of using GET to develop transgenic animal models of human disease are based on the Western allopathic medical tradition. This is by no means the only medical paradigm, and it has a monopoly neither on truth nor on ways of preventing and reversing human diseases. Nor is it, as biomedical science claims to be, purely objective and value free.

Much of the research on developing animal models of human disease using GET is aimed at finding new drugs and other treatments for such common Western diseases as arteriosclerosis, high blood pressure, stroke, diabetes, cancer, obesity, Alzheimer's disease, and kidney and liver failure. Yet many of these diseases could be prevented and even reversed by people eating little or no animal fat or animal protein and more organic and unrefined cereals, fruits, and vegetables. Instead, we are moving toward social acceptance of xenotransplantation, with the wholesale production of transgenic pigs (expressing complement masking proteins) with organ parts for human patients. It is unlikely, however, that such transplants will ever become a reality or offer patients any significant quality of life.

This is because of potential human health risks from animal viruses and prions, increased societal costs, and other competing medical alternatives and public health priorities.

Furthermore, of the 2,500 known human genetic diseases, only 400 are caused by a single gene defect. Most human diseases are multigenic and involve environmental causal factors (e.g., dioxins, radiation), as well. These are generally too complex to be modeled in animal systems. Thus, the use of GET to create specific mouse models of human diseases in order to develop new drugs is limited to this smaller number, which represents about 1 percent of all human illnesses.*

Certainly genetic-engineering biotechnology will have some success in the diagnosis and treatment of certain diseases. But identifying the original causes of most diseases, and therefore their prevention, does not lie in a full understanding of human genetics and a completed human genome map. The creation of transgenic animal "models" of human diseases and the promises of medical genetic breakthroughs are more hype than they are scientifically valid.[14] Similarly, a reliance on genetic engineering to intensify agricultural, aquacultural, and agroforestry production will not be sustainable because those

---

*Nevertheless, the number of transgenic animals used in British laboratories increased to over 300,000 in 1997, a sevenfold increase since 1990, and far more than the numbers involved in cosmetics, alcohol, and tobacco research—and not counting transgenic fish. The number of animal experiments is rising, said to be due to the rapid expansion of genetic research (The BBC Online, May 5, 1999). A similar increase has occurred in the United States, but exact figures are not available. U.S. researchers insisted in the 1980s that creating transgenic mice as more accurate disease models of human disorders would mean fewer animals would be used, while in fact, more animals are used than ever before.

industrialized systems of production are ecologically unsound to begin with. Genetic engineering is no substitute for sound ecological methods and management practices.

Turning farm animals into "bioreactors" to produce pharmaceutical products, some of which will be used to treat humans afflicted with genetic diseases, is also questionable in terms of possible public health risks, treatment costs, resource allocation, and competing social priorities in the health-care field. We need to ask how many millions of surrogate transgenic animals might we be using a century hence if we become medically dependent upon them now. Will we suffer any more or any less?

What will be left of biodiversity and of nature if we continue to multiply and use GET not to help control our own numbers but to increase the number of farm animals for our consumption? In sum, a number of complex ethical and scientific questions and concerns have yet to be addressed with respect to biomedical research and practice, conventional industrialized agriculture, and other commercial applications of genetic-engineering technology, from bioremediation to genetic screening and discrimination (that is, denying people with certain gene anomalies a job or insurance).[15] Without such consideration, the misapplication of GET will be difficult to avoid and the risks and costs may far outweigh the hoped-for benefits.

Being aware of these and other potentially harmful consequences of cloning and genetic-engineering biotechnology in general, animal protection organizations have urged the president's National Bioethics Commission to consider animal well-being in its evaluation of all cloning applications and to put on the commission a bioethicist who is informed about animal sentience, welfare, and behavior.

We shall see. As I left the Senate hearings on cloning, where I lamented in uncharacteristic silence the pervasive anthropocentrism and pro-biotechnology tenor of the entire proceeding, I met a science reporter who told me he was deeply disturbed about the creation of genetically engineered animals and said he was aghast at people "creating transgenic animals to maintain our depraved human existence."

We have become depraved, I believe, by our own seduction into the kind of thinking Senator Bill Frist endorsed, quoting the "father" of modern medicine, Dr. William Osler, in his opening statement at the hearings: "To wrest from nature the secrets which have perplexed philosophers in all ages, to track to their sources the causes of disease. . . . These are our ambitions." Our anthropocentrism has disabled us from "tracking" at least half or more of the causes of disease that arise from us humans rather than from nature. We still blame nature for floods, famines, and pestilence, yet most of these, like global warming and *E. coli* food poisoning from hamburgers, are anthropogenic.

Dr. Osler reflects the flawed worldview of the dominant culture of bioindustrialism, which has been obsessed with wresting from nature the dream of Francis Bacon, the seventeenth-century founder of industrialism and prophet of biotechnology —namely, total control over life and the creative process. Perhaps Dr. Osler meant well, but his worldview is as limited as the president's present Bioethics Advisory Commission. Animals are not on the agenda of human concern. Until they are, cloning and other developments in biotechnology are likely to do more harm than good and to serve the interests of an increasingly depraved human existence.

It was evident from the congressional hearings that the establishment is comfortable playing God with animals but not with ourselves. The bioethical questions about how far we can and should go in interfering with nature, with life and earth's creation, have yet to be openly and fairly addressed.

POSTSCRIPT

In an unprecedented move in February 1998, the Dutch minister of agriculture put a stop to cloning experiments being done by Pharming, a company based in the Netherlands that specializes in producing drugs in milk. According to the government, Pharming must desist from cloning cows until it proves that drugs from such animals are better than those made by other methods. The ban came a few hours after Pharming announced the births of Holly and Belle—two calves cloned from embryonic cells—and applies only to work at Pharming. In response, the company announced plans to move its cloning research to the United States.

CHAPTER 6

How Prevalent—and Safe— Are Genetically Engineered Foods?

> We may end up with prescription foods or prescription diets. We may have Monsanto diet No. 3, for example.
>
> —Robert Shapiro,
> Monsanto chairman and CEO

> Without labeling, it will be very difficult for scientists to trace the source of new illness caused by genetically engineered food.
> —Dr. John Fagan

In its issue of May 2, 1997, the *New York Times* (p. B8) reported having hired Genetic ID,* a company that has developed the technology to test food for genetically engineered food ingredients, to test soy-based baby formulas and eight other products made with soy or corn. The formulas—Carnation Alsoy, Similac, Neocare, Isomil, and Enfamil Prosobee—all tested positive;

---

*500 North Third Street, Suite 208, Fairfield, IA 52556.

Eden soy milk tested negative. Morningstar Farms Breakfast Links and Morningstar Farms Better 'n Burgers, Betty Crocker Bac-os Bacon Bits, all soy-based products, turned up positive, as did some corn-based chips—Tostitos Crispy Rounds, Doritos Nacho Cheesier, and Fritos.

Genetic ID provided the following list of genetically engineered foods that have been approved by the FDA or are in the pipeline for approval; an asterisk indicates foods currently in the marketplace.

Alfalfa	Lettuce
Apples	Papayas
Asparagus	Peanuts
Barley	Pepper
Beets	Potatoes*
Broccoli	Raspberries
Canola (rapeseed oil)*	Rice
Carrots	Salmon*
Cauliflower	Soybeans*
Cheesemaking enzymes	Squash
(chymosin)*	Strawberries
Chestnuts	Sugarcane
Corn*	Sunflowers
Cotton (cottonseed oil)*	Tomatoes*
Cucumbers	Walnuts
Flaxseed	Watermelons
Grapes	Wheat
Kiwi fruit	

In the summer of 1998 the Union of Concerned Scientists published (in its periodical, *The Gene Exchange*) a dinner menu that is made up of currently marketed transgenic foods in order to illustrate the fact that such foods are being consumed already by a totally uninformed public (see pages 126–27).

NUTRACEUTICALS: DESIGNER "MEDIFOODS"

There is much industry hype over genetically engineering crops to be more nutritious and of medical benefit, the line between food and medicine becoming blurred in the process. New designer "medifoods" are on the drawing boards touting health benefits such as lowering cholesterol, providing more of certain amino acids and other essential nutrients, and helping boost the human immune system. They will compete with conventional nutrient supplements, herbal medicines, and natural foods such as olive and flax-seed oils and soybeans, which contain beneficial nutraceuticals that help lower blood cholesterol and fight cancer.

DuPont is engineering canola and soybeans to eliminate trans-fatty acids. Monsanto is developing a cholesterol-lowering corn-fiber oil. Monsanto's GE canola oil, high in laurate, a fatty acid, is being marketed as a replacement for cocoa butter in chocolate and as a dairy replacement in other products. Algae and fungi are being engineered to produce omega-3 fatty acids, which boost cardiovascular health, as an alternative to fish oil.

Transgenic "pharm" animals are producing nutraceuticals in their milk—such as prolactin from goats, which purportedly enhances the human immune system; lactoferrin from cows,

The Transgenic Café

Appetizers

Corn tortilla chips (firefly gene)
or *potato chips* (chicken gene)
Salsa of tomato (flounder gene),
pepper (virus gene), and
onion (garlic gene)

Soup and Salad

Creamy soup of broccoli
(bacterial gene)
Garden salad of lettuce (tobacco gene),
radicchio (bacterial gene),
and cucumber (tobacco gene)

Entrées

Catfish (trout gene)
fried in oil pressed from canola
(California bay gene)
Grilled veggie burger of soybean
(petunia gene) and carrot (petunia
gene)—served with roasted potatoes
(waxmoth gene) or eggplant (bacterial gene)
and bread sticks of wheat (bacterial gene)

Dinner Menu*

Cheese

Assorted cheeses
(engineered bacterial rennet)

Dessert

Fresh fruit plate of papaya (virus gene),
strawberries (undisclosed gene),
raspberries (virus gene), and melon
(tobacco gene)—sprinkled with
sunflower seeds (cowpea gene) and
walnuts (barley gene)—served with cream from
engineered bovine growth hormone (BGH)–injected
cows and sticks of sugarcane (bacterial gene)

Beverages

Chilled apple juice (silk moth gene)
Porter beer of barley
(undisclosed gene)
White wine of grapes (virus gene)

* The menu includes only some of the many engineered food organisms
approved for field testing or commercialization in the United States.
The source of the foreign gene, where available in government documents,
is in parentheses. For more information, see USDA and FDA web sites at
aphisweb.aphis.usda.gov/bbep/bp/index.html
and vm.cfsan.fda.gov/ ~lrd/biocon.html.

which has anti-inflammatory and antibacterial properties; and bile salt stimulated lipase from sheep, an enzyme that may benefit victims of cystic fibrosis as well as premature infants.

The market in bioengineered nutraceuticals will soon be competing with conventional nutraceutical, nutrient-supplement, and pharmaceutical products and markets. There may be many benefits from new products from genetically engineered algae and fungi (provided there are no "escapes" into the natural environment), but there never will be any safer and more reliable natural "health foods" than those that have been produced organically, and no better way for an informed public to eat a varied diet, primarily of plant origin, that is certified as organic.

Hendrick Verfaille, president of Monsanto, told a 1998 press conference that his company is developing a variety of wonder drugs and superfoods from genetically engineered plants, for which Monsanto hopes to get government approval in the next three years.[1] Monsanto is researching a range of new foods that are aimed at counteracting diseases like colon cancer and diabetes. Verfaille proclaimed, "We believe people are going to be able to eat themselves healthier."

Employing genetic engineering to make new nutraceuticals raises many questions. How pure and safe are those derived from farm animals' milk and possibly, with new developments, from their urine? Are other essential nutrients lowered in plants that have been engineered to produce more of some particular protein? Will whole foods and natural supplements be pushed off the market shelves by competition and monopoly, and consumers be obliged to accept a fragmented diet of nutrient-deficient foods bolstered by all kinds of genetically engineered nutraceuticals (GENS)?

We should reflect on the need to create new Frankenfoods and GENS. As I have emphasized in another book,[2] we would have little need for nutraceuticals if our crops were raised organically and came from soils that had not been depleted of essential nutrients, as is the case with and consequence of conventional industrial agriculture. We must also consider the ethics of producing nutraceuticals and various health-care products from animals and the steps that were taken in creating these "pharm" animals. For example, we may soon have sheep, goats, and cows producing calcitonin, a peptide used in the treatment of osteoporosis and other medical conditions, in their milk. A first step was to splice the calcitonin gene from salmon into rabbits and to demonstrate that it can be harvested from the milk of those rabbits.[3]

In September 1998, the European Union ruled that food manufacturers must state on a label when genetically modified ingredients have been used, but consumer groups pointed out a major loophole. The regulations exempt foods containing soy oil and soy derivatives such as lecithin, which accounts for 90 percent of soy products. More than 60 percent of processed foods contain soya—from sausage skins to hypo-allergenic milk, from bread to chocolate, from diet foods to baby foods—and more than half the soya in British products is expected to be genetically modified within two years.

As ethics professor Paul B. Thompson emphasizes, changes in our food that are wrought by new technologies cannot be determined to be in any way unnatural or publicly unacceptable by policy makers if the primary, if not the sole, criterion of acceptability is food safety.[4] This procedural issue of by whom and by what criteria food should be labeled is a major bioethical issue. Substantial public concerns—religious, ethical, and aesthetic (e.g., desiring whole and natural foods)—should not be discounted by policy makers, who ought not to violate liberties of conscience, consumer sovereignty, and the right of informed choice. I do not entirely agree with Thompson's proposal for a "no-biotechnology" type of label for people desirous of access to whole foods (and beverages), since it would leave unlabeled all genetically engineered processed and prepared foods. Both Thompson and I contend that the opposition to labeling GE food by the food industry and many scientists is based on specious claims about the safety of such food. This kind of public policy, touted as objective and "science based," discounts bioethical concerns and usurps consumer sovereignty.

According to *The Gene Exchange* (Fall/Winter 1998), the world's farm acreage sown to genetically engineered crops has burgeoned to 70 million acres in 1998 from only 5 million acres two years earlier; 51.3 million acres were used in the United States—predominantly for soybeans, corn, cotton, canola, and potatoes.

Close to one-quarter of Canada's canola (much of which is imported to the United States) is genetically engineered. U.S.

agricultural biotech product sales totaled some $520 million in 1998, averaging an annual growth rate of 22 percent (*Illinois Agrinews*, May 8, 1998). Human biotech-based therapeutics dominate the genetic-engineering industry with an estimated $9 billion in sales in 1998 and an additional $2 billion in diagnostics, including genetic screening and disease-detection agents.

The *New York Times* of July 20, 1998, reported the following:

> On January 1, 1998, the U.S. government gave the green light to genetically modified soybeans, cotton, corn, summer squash, potatoes, canola oil, radicchio, papayas, and tomatoes. The amount of genetically modified soybeans, cotton, and corn on the market is significant. According to one study, the gene-altered corn crop in the U.S. this summer is estimated to be 32 percent of the total, for soybeans 38 percent, and from Canada 58 percent for canola oil.

As ocean supplies of seafoods decline because of overfishing and pollution, aquaculture—farming shellfish and finfish—has increased such that over 25 percent of all fish consumed worldwide are now farmed. Many species are being genetically engineered to speed growth, alter flesh quality, and increase cold and disease resistance. Analyst Dr. Rebecca Goldburg reports that over twenty species have been genetically engineered (see table on page 132).[5] She reports:

> Concerns about the potential ecological impacts of transgenic fish are especially serious—for two reasons. First, aquaculture facilities are not escape proof—sometimes large numbers of

fish are accidentally released. For example, last summer 300,000 Atlantic salmon escaped from a single Washington State salmon farm. Escapes from Washington State and British Columbia salmon farms are so common that it is no longer remarkable to catch Atlantic salmon in Pacific Northwest waters. Second, unlike many domesticated terrestrial farm animals and crops, most farmed fish have not been debilitated by generations of selective breeding. As a result, transgenic fish may easily survive and reproduce after escaping an aquaculture facility—posing, on the whole, more potential for ecological damage than transgenic crop plants and livestock.

Escapes of transgenic fish are cause for concern, since they could travel vast distances, spread new diseases, out-compete and hybridize with native fish, and decimate wild fish stocks.

Fish Species That Have Been Genetically Engineered

Abalone	Medaka
Atlantic salmon	Mud carp
Bluntnose bream	Northern pike
Channel catfish	Penaeid shrimp
Coho salmon	Rainbow trout
Common carp	Sea bream
Gilthead bream	Striped bass
Goldfish	Tilapia
Killifish	Walleye
Largemouth bass	Zebrafish
Loach	

PUBLIC OPPOSITION INTENSIFIES

According to Professor John Beringer, "The key issue for biotechnology is how to create desire. . . . The battle for people's minds cannot be based on explanations. It must be based on lust." Increasing public concerns over the risks of genetically engineered crops and foods have led five of the European Union member states—Austria, France, Greece, Luxembourg, and the United Kingdom—to impose either specific bans or some form of moratorium on genetically modified (GM) plants. As of November 1998:

➤ Austria and Luxembourg still have their long-standing ban on Novartis GM maize in place, despite all the efforts of the European Commission to persuade them to lift the ban.

➤ France has imposed (since July 1998) a moratorium on all GM plants with indigenous relatives, such as oilseed rape and beet, for a two-year period pending further research.

➤ Greece banned AgrEvo's genetically modified oilseed rape in the spring of 1998.

➤ The United Kingdom's government just announced a three-year moratorium on insect-resistant GE crops, as well as a program of restrictions for commercial release of herbicide-tolerant crops.*

➤ The European Parliament's Environment Committee has also called for moratorium on new authorizations of genetically modified organisms for commercial purposes while the scientific issues are being clarified.[6]

---

*In May 1999, three giant multinational food companies—Unilever UK, Nestle, and Cadbury—announced that they will no longer market GE foods or ingredients in the United Kingdom (*The Independent*, London, May 2, 1999).

➤ In early 1999 the Australia–New Zealand Food Standards Council passed a requirement that all food containing genetically modified material be labeled.

American consumers are becoming concerned about the socioeconomic impacts of biotechnology on farmers and rural communities. A 1992 interim report on a survey by the USDA's Extension Service revealed some significant public perceptions and attitudes:[7]

➤ 85 percent of those surveyed feel it is important to label foods if biotechnology is used.
➤ 94 percent want to know if pesticides are used.
➤ 88 percent want irradiated foods to be so labeled.
➤ 24 percent feel that the use of biotechnology to alter plants is morally wrong.
➤ 53 percent feel it is morally wrong to alter animals, and expressed more concern over eating meat and dairy products developed with biotechnology than eating genetically engineered fruits and vegetables.
➤ Consumer acceptance of plant-to-plant genetic engineering is 66 percent; however, approval fell to 39 percent for animal-to-animal genetic engineering, and to 10 percent for human-to-animal transgenic alteration.

A European Commission survey of people's views about genetic engineering was conducted throughout the whole of the EU and found that:[8]

➤ People feel that the biotechnology developments that pose the greatest risk are the introduction of human genes into animals to produce organs for transplants, and the use of modern biotechnology in food production.

➤ Fewer than one in four people think that current regulations are sufficient to protect people from the risks of modern biotechnology.

➤ Only a minority of those surveyed found genetic manipulations with a view to producing organs for transplants, and the development of genetically modified animals for health-related research to be morally acceptable.

In Europe, people are arguably more politically involved than in the United States with the global socioeconomic and environmental costs and risks of agribiotechnology and the ethical ramifications and medical hazards of xenotransplants, genetically engineered vaccines, and "pharm" animal health-care products. In the fall of 1997, five naked activists scaled the office building of Bartle Bogle Hegarty in England to protest this ad agency's work for Monsanto to give GE food an image makeover. A strong European coalition of Greenpeace, the Gaia Foundation, and Friends of the Earth have formed Reclaim the Streets, a resistance movement against monoculture, genetic imperialism, capitalistic industrialism, and the annihilation of biological and cultural diversity and autonomy.

Prince Charles, an outspoken opponent of agribiotechnology being adopted by U.K. farmers, contends that it "takes mankind into realms that belong to God and to God alone."

Some transgenic test crops grown for the first time in the United Kingdom were destroyed by activists in 1998, and an organic farmer filed suit against Monsanto, claiming risk to his livelihood from genetic pollution of his crops and seed stock from neighboring fields where genetically engineered canola (or rape) was being grown.

Monsanto not only spent a purported $1.6 billion on public relations and advertising to get its genetically engineered product publicly accepted in the United Kingdom and Europe in the spring of 1998, but also lined up leaders of several African countries to endorse genetic engineering as the way to end world hunger in the next century.

According to reporter Zadie Neufville (Euro News Service, London, August 3, 1998), the Monsanto "Let the Harvest Begin" ad campaign included the following statement:

> We all share the same planet—and the same needs. In agriculture, many of our needs have an ally in biotechnology and the promising advances it offers for our future. Healthier, more abundant food. Less expensive crops. Reduced reliance on pesticides and fossil fuels. A cleaner environment. With these advances, we prosper; without them, we cannot thrive.
> . . . Biotechnology is one of tomorrow's tools in our hands today. Slowing its acceptance is a luxury our hungry world cannot afford.

She further reported that:

> A counter-attack has been mounted. A media alert distributed by the Panos Institute of London, a non-governmental organi-

zation that works to stimulate debate on global environment and development issues, said [Friday] that senior African politicians, scientists and agriculturists have released counter-statements to the European press. African delegates to the United Nations Food and Agriculture Organization (FAO) in a counter-attack on the planned Monsanto ads said in a joint statement, "We strongly object that the image of the poor and hungry from our countries is being used by giant multinational corporations to push a technology that is neither safe, environmentally friendly, nor economically beneficial to us." The delegates, who included representatives from all African nations in the UN except South Africa, accused Monsanto of "threatening and jailing" U.S. farmers who save seeds for planting the next year's crop. The delegates denounced Monsanto's interest in the environment. "Its major focus is not to protect the environment, but to develop crops that can resist higher doses of its best-selling chemical weed killer Roundup."

Dr. John Fagan, an American biologist and a spokesman for Consumers International—a coalition of more than two hundred consumers' groups—says that as much as 70 percent of U.S. packaged food contains genetically engineered ingredients such as soy and canola oil. The United States and Canada are major producers of soy and canola. Dr. Fagan says genetically altered food should be so labeled because the long-term health effects are not properly understood.

Mainstream opponents of biotechnology are opposing more than the technology itself. They are opposing a political system that permits genetic piracy, genetic capitalism, and genetic monopoly by multinational agricultural and pharmaceutical life-science industries. They see biotechnology as having some potential benefits but argue that those benefits will never be realized as long as profits come first and the market-driven thrust of the food and drug industrial complex results in widespread misapplications of genetic engineering. In the first place, these profit-making pressures tend to displace conventional crops, organic agriculture, and natural foods in a global game of monopoly. Second, the profit motive leads to the marketing of medical products, many derived from transgenic animals, to treat, rather than prevent, a variety of human diseases that have a dysfunctional genetic, neuro-endocrine, or immune system basis. The vast capital investment and overemphasis on genetic engineering as a panacea for such maladies undermines a more holistic approach to human health-care maintenance and disease prevention. This necessitates a much more stringent prohibition of industrial pollutants, the adoption of organic agriculture, and decisions by consumers to eat with conscience and assume greater responsibility for their own health.[9]

Agricultural applications of genetic engineering, like the conventional industrial agriculture of which it is an extension, threaten to disrupt the self-organizing, self-replicating, self-healing, regenerative, and self-sustaining capacities of organisms, ecosystems, communities, and national economies. As Vandana Shiva notes, "Ecological problems arise from applying the engineering paradigm to life."[10] The deformities of cloned animals and the genetic pollution by GE crops are indicative of our lack of understanding of natural systems and processes.

Shiva sees that "when organisms are treated as if they are machines, an ethical shift takes place—life is seen as having instrumental rather than intrinsic value." Genetic engineering is now part of the globalization of an industrialized, market-driven economy that sees life as a mere commodity. Via intellectual property rights (patents on genetically engineered products) and the World Trade Organization's legal system and conventions, a global monoculture is being created that is displacing nature and indigenous cultures. As Shiva points out, "GATT is the platform where the capitalistic, patriarchal notion of freedom as the unrestrained right of men with economic power to own, control, and destroy life is articulated as free trade." The globalization of this industrial monoculture undermines local economies, self-governance, and self-determination. As Shiva rightly concludes, "An intolerance of diversity is the biggest threat to peace in our times; conversely, the cultivation of diversity is the most significant contribution to peace—peace with nature and between diverse peoples."

Resistance to biotechnology in the third world is intensifying as such countries become the victims of "biopiracy"—the corporate theft of indigenous knowledge and genetic resources, as exemplified by W.R. Grace Company of the United States claiming patents on India's traditional neem-tree products on the basis of Grace having modernized extraction methods. Genetic piracy is not a figment of the imagination of anti-biotech extremists. There was an international outcry when two Australian government agencies sought to obtain patents on two species of chickpeas developed by subsistence farmers in Iran and India.[11] The seeds were obtained in Syria from one of eleven gene banks around the world supervised by the United Nations' Food and Agriculture Organization in Rome and the Consultative Group on International Agricultural Research, an intergovernmental agency based at the World Bank in Washington, D.C.

In a devious ploy to legitimize genetic piracy, life-science multinationals are pressuring developing countries to put world patents on all their valuable native seeds and traditional medicines. This process of protecting "intellectual property rights" is an extremely costly process. If it is not completed by the arbitrary deadline of 2010, and if the deadline is not extended by the WTO, these multinationals are claiming that they will have free license to plunder the wealth of biodiversity in these poorer countries and to put patents on whatever they want.

One way to control agriculture is to patent a process or product and then do nothing with it, effectively precluding others from developing the process for themselves. For example, Australian agronomist Richard Jefferson has described how some wild plants can produce seeds without sex, a process called *apomixis*.[12] If this genetic trait were transferred to crops, he suggests, farmers in developing countries would not have to purchase new hybrid seed every year. Soon after the feasibility of this process was demonstrated, however, the *Biotech Reporter* (November 1997, p. 14) announced that World Patent number 9710704 had been granted to the U.S. secretary of agriculture for "apomixis for producing true-breeding plant progenies." By securing a patent, this technology can be controlled or suppressed, whichever would be in the best interests of the multinational biotech companies that want farmers to buy their superseeds every year.

The collective consequences of the ways in which genetic engineering technology is being misapplied is one of many human influences, including overpopulation and overconsumption, that will soon mean a world devoid of a whole earth, of a nature that is vibrant and as rich in biological diversity as the world is enriched by our own cultural diversity.

The "anti-monoculture resistance movement" is now spreading worldwide as diverse people come together, ever more effectively via the Internet. They see biotechnology as a growing menace to biocultural diversity and abhor the transformation of the natural world into a bioindustrialized monoculture of productive efficiency, driven by consumer demand and corporate profits and growth. The growth and diversification of multinational life-science corporations is coupled with the shrinking of the natural world and the irretrievable loss of biological diversity. Some British scientists see this loss as a natural result of human evolution and suggest that, provided the major family branches on the evolutionary tree of life are not severed, but only twigs cut off (i.e., individual species), other twigs will eventually sprout (evolve) as they always have done, so species extinction is not going to "make any difference to us personally." So says Oxford ecologist Dr. Sean Nee, co-author with the U.K. government's chief scientist, Professor Robert May, of an article in *Science* that says, "Approximately 80 percent of the underlying tree of life can survive even when approximately 95 percent of species are lost." And, "We have shown that much of the tree can survive vigorous pruning."[13] While Dr. Nee recognizes that each species is unique and that extinction eliminates it for eternity, he told a news reporter, "I guess I'm taking the naughty side—it's a bit more fun" (*The Independent*, October 23, 1997).

The reductionistic view of these scientific "experts" is flawed in many ways by their faith in the central dogma that genes alone *control* life. First, they do not see that what sustains the tree of life is the tree itself, each species contributing to the functional integrity of the earth's ecology. Second, they ignore the probability that mass extinctions could cause great harm to the tree of life and harm us also, since we are dependent on the tree, as we are a small, possibly diseased, growth on one of its many limbs.

Proper food labeling may slow down the biotech explosion, but what is really needed is a moratorium on the planting of all genetically engineered crops and on the release of any and all genetically engineered organisms into the environment. This moratorium would end on a case-by-case basis as each crop or organism is shown to cause no genetic pollution or harm to soil microorganisms or other natural fauna or flora that are currently ecologically beneficial and part of wild nature.

GE CROPS=INFECTED FOOD

The Bt insect toxin is similar to the insecticidal lectins in transgenic potatoes. These novel genetic traits are put into these crops in combination with a virus—the cabbage mosaic virus—that has been bioengineered to act as a carrier and promoter for the transgenic traits. The lectin genes in these new super-potatoes come from the snowdrop flower. Snowdrop lectins have also been engineered into transgenic rice, oilseed rape (canola), and cabbages. The patent to this cauliflower mosaic promoter *that is used in most genetically engineered foods available worldwide* is actually owned by the UK's Minister of Science, the billionaire supermarket tycoon Lord Sainsbury of Turville.[14]

Increasing public concern over the safety of genetically engineered Frankenstein foods and over the commitment of government to effectively regulate these new foods came to a head in the United Kingdom in early 1999.[15] Apparently, the research findings of a reputable scientist, Dr. Arpal Pusztai from the government's own Rowett Research Institute, revealed adverse health effects in rats fed genetically engineered potatoes. This resulted in suppression of his research, a gag order, and forced early retirement. He reported abnormal organ development and weakening of the immune system attributable to either the cauliflower mosaic virus that was used as a promoter, or to the active genetically spliced insecticidal lectins present in these GE potatoes. Liver, brain, and heart sizes of the rats decreased. Subsequent studies by another scientist confirmed Dr. Pusztai's findings. The real problem for the biotech food industry is that the mosaic virus has already been used in the modified tomato paste, soya oils, and maize that the U.K. government and the European Union approved for use in industrial and convenience foods and which are now in hundreds of products on supermarket shelves.

Given that alien genes, viruses, and bacteria are being put into our food by the life-science industry, to simply call this "genetically modified" or "genetically enhanced" food is deliberate public misinformation. Such modified foods would be better called *"infected foods,"* because that is essentially what they are, infected indeed with viral vectors and promoters, bacterial pesticides and insect toxins. We know nothing about the health and environmental risks of these infected foods. But we do know that introduced viruses can combine with endogenous, naturally occurring virus particles in plants' cells, the consequences of which could be extremely harmful.[16]

DNA released from living and dead cells can persist in the environment and be transferred in a variety of ways to other living organisms. [17]

CHAPTER 7

Biotechnology and Genetic Pollution

We have taken a successful and
extremely useful theory and paradigm
of the gene and have illegitimately
extended it as a paradigm of life.

—Professor Richard C. Strohman

A global alliance of individuals and organizations is most urgently needed to prevent the first creation—the natural world, wildlife and wildlands, natural foods, and humanity— from being obliterated by the monocultures of global industrialism and consumerism. The natural world is being rapidly supplanted by the second creation, as humankind plays God with DNA and remakes the first creation in the human's own projected image. Even if the current rate of habitat reduction and loss of natural biodiversity is somehow stopped and wildlife and wilderness areas are protected from further encroachment, fragmentation, and species depletion, these areas will still not be safe. Genetic pollution will be almost inevitable. Genetically engineered crops and other patented life forms may soon have spread their anomalous transgenes into the earth's life-stream through transfer, via pollen and other means, into the germ plasm of nature and natural species of many life forms—plant, animal, bacterial—in many parts of the world. It may already be too late to reverse this process because there is already genetic pollution in many parts of the world, afflicting many unique ecosystems.

The U.S. agribiotech industry is now so deregulated that a company planning to do field tests on a new GEO simply has to send a letter of notification to the USDA. Similarly, a company has only to submit "summary information about the safety and nutritional assessment of the plant [created by genetic engineering] to the FDA" (FDA *Veterinarian*, January/February 1997, p. 12). This "fast-track" approval process essentially gives biotech companies a free hand in developing and marketing GEOs without any detailed environmental or socioeconomic impact assessments being conducted.

The impact of transgenic crops and genetically engineered microorganisms and insects released in developing countries raises concerns.[1] Developing countries are still centers of biodiversity and are important for the future of the world's major food crops. The U.S. National Academy of Science's 1989 report *Field Testing Genetically Modified Organisms: Framework for Decisions* warns, "The incidence of hybridization between genetically modified crops and wild relatives can be expected to be lower [in the U.S.] than in Asia Minor, Southeast Asia, the Indian subcontinent and South America, and greater care may be needed in the introduction of genetically modified crops in those regions."

According to Andre de Kathen, by 1996 there had been some 159 releases of genetically engineered crops, such as maize, cotton, and soya bean, with herbicide, insect, and virus resistance, in such countries as Argentina (43), Cuba (13), Puerto Rico (21), and Mexico (20).[2] Foundation stock of GE seeds, like Dekalb's Roundup Ready–resistant corn (being marketed under a cross-licensing agreement with Monsanto), have been produced by growers in Hawaii, Argentina, and Chile.

The loss of genetic diversity via genetic pollution by GE crops is a major concern; the first known instance of genetic pollution was reported in 1996 by Danish scientists who found that herbicide-tolerant genes from engineered oilseed rape became established among related "weedy" forms of rape after just two generations of inbreeding.[3] Soon after this report, Dr. Norman Ellstrand and his co-workers at the University of California at Riverside confirmed that via simple pollination, GE sorghum (one of the world's most important cereal crops) had transferred an herbicide resistance to Johnson grass, one of the world's most troublesome weeds.

The cross-contamination of conventional crops, like that of conventional flax by GE flax high in linoleic acid, for example—which would be especially easy if they are grown close by—could result in harm to both varieties and render both unfit for their intended purposes. Weedy relatives of crops like GE celery, asparagus, carrots, and sunflowers are other examples of plants that could be cross-contaminated. If they are cross-contaminated with genes that produce insecticidal toxins in the plant's cells, then all kinds of insect, reptile, amphibian, avian, and small-mammalian life are going to be wiped out as the insect part of the natural food-chain is exterminated.

A concern that genetically engineered crops that produce their own insecticide, like Bt, could harm beneficial insects that help control crop pests has been verified by a team of scientists led by Dr. Angelika Hillbeck (*Environmental Entomology* 27:480, 1998). When insect pests that had been eating Bt corn were fed to larvae of the green lacewing, a farmers' friend and a major predator of corn pests, their mortality rate nearly doubled and their development was delayed. The Scottish Crop Research Institute found that ladybugs had reproductive problems and lived shorter lives after consuming aphids that had been eating transgenic potatoes producing insecticidal lectins (from a snow-drop plant gene).

Errors in genetic engineering also occur, as illustrated by Monsanto's May 1997 recall of "small quantities" of genetically engineered canola seeds containing an unapproved gene that got into the product by mistake. Sufficient seed for 600,000–750,000 acres of land was recalled.[4]

Other problems with genetically engineered crops have come to light. Since its commercialization, Monsanto's Bt cotton failed to control bollworms in field tests from Texas to Georgia and probably accelerated their development of resistance to Bt.[5]

Gene Exchange, a publication of the Union of Concerned Scientists, has noted that the EPA does not evaluate the health and environmental risks of genetically engineered herbicide-tolerant crops. Particularly disturbing is a report that Monsanto's Roundup Ready–resistant GE soybeans have higher than normal levels of estrogen.[6] Earlier consumer risks of GE food were reported by J. Nordlee et al., who found that an allergen from Brazil nuts retained its allergenic properties when transferred to soybeans.[7]

The widely reported discovery in 1996 of wild ryegrass on a farm in northern Victoria, Australia, developing herbicide resistance after exposure to Roundup, used in many countries as a component of conservation or "no-till" farming, provides another example of the unplanned effects of GE seeds that are herbicide tolerant.[8]

After gene-transfer between wild, cultivated, and weed varieties of the sugar beet was demonstrated in France, researchers in the United Kingdom have also raised the flag on the risks of transgenic crops transferring genes to close relatives growing nearby.[9] These concerns are heightened by other research demonstrating that viral RNA or DNA, inserted into a plant to make it virus resistant, may recombine with genetic material from an invading virus to form new, more virulent strains.[10]

At a meeting of the Ecological Society of America in Baltimore on August 6, 1998, Dr. Allison Snow of Ohio State University detailed how she and a Danish scientist had found new evidence that genes can spread from herbicide-resistant transgenic crops to weeds, making the weeds stronger than ever.[11] In the worst-case scenario, "super weeds" spread transgenic traits via their pollen to more of their own kind and also to normal crop relatives that have not been genetically engineered.

These concerns are further validated by the findings of Dr. Joy Bergelson of the University of Chicago, who has shown that plants genetically engineered to resist the herbicide chlorsulphuron were able to fertilize other plants at a rate twenty times greater than plants that had been mutated.[12]

There is also concern that some genetically engineered crops may be genetically unstable and have lower viability as a consequence of alterations that disrupt their normal physiology and place an additional metabolic burden on the plants.

Another warning of the environmental risks of GE organisms comes from the research of Elaine Ingham and Michael Holmes, who found that the soil bacterium *Klebsiella planticola*, engineered to produce ethanol, reduced the mycorrhizal fungus population in the roots of wheat plants, which died as a consequence.[13] The widespread use of antibiotic-resistant genes as identification tags in GE seeds poses a risk to soil microorganisms and the possible transfer of resistance to intestinal bacteria in livestock and people who consume grains and other GE foods.

A study by the State Institute of Ecology for the Ministry of the Environment of Niedersachen, Germany, demonstrated the "escape" of genetic material from a test plot of rapeseed, confirming research results from France and Denmark.[14]

More instances of genetic pollution were reported in 1998: canola resistant to Roundup herbicide turned up in northern Alberta; GE maize cross-pollinated an adjacent field in Germany; and jointed goat grass acquired resistance to imazamox herbicide in the United States from wheat engineered to be resistant to it.

Millions of acres of genetically engineered crops are now being grown around the world, the pollen of which, carrying new genes, cannot be contained. This will mean genetic pollution of natural crop varieties and of wild plant relatives.* Genetic pollution will also result following the deliberate or accidental release of genetically engineered bacteria, insects, fish, and other life forms into the environment. Unlike other forms of pollution, genetic pollution is uncontrollable, irreversible, and permanent, thus posing a major threat to biodiversity and to the bio-integrity of the entire life community. Genetic pollution by "Terminator" and "Verminator" gene technology (see chapter 2) could result in widespread sterility and other problems in conventional crops and wild plants.

According to *The Economist* (July 25, 1998, p.77), the world's $400 billion timber industry is investing heavily in artificially enhancing production through genetic engineering, though many ecologists are concerned about the potential for genetic pollution of remaining natural forest ecosystems.

Several research centers are now genetically engineering trees to make them grow faster and resist diseases and salinity. The *Gene Exchange* (Summer 1998, p.11) reports:

---

*The U.S. Prima Terra food company lost $170,000 in sales after its corn chips exported to Holland failed a random "gene scan" by government inspectors. Pollen from GE corn apparently had blown into an organic farmer's cornfield in Texas (*Wisconsin State Journal*, March 24, 1999).

Scientists at Union Camp, Westvaco, and other paper companies are engineering sweet gum, paulownia, cottonwood,
among other trees, hoping to create a supertree—one that
grows faster than normal but retains hardiness. So far, they
have gotten rapid growth and delicateness—tall, fast-growing
specimens that require special treatments like fertilizing,
pruning, and weed control. Back to the drawing board.

Other scientists have had more success engineering trees
to control weeds and insects that plague tree plantations.
The Oregon State University Tree Genetic Engineering
Research Cooperative—a consortium of companies, government agencies, and universities—has engineered hybrid
poplars to resist the herbicide glyphosate and produce insecticidal Bt toxins. Glyphosate, which is toxic to ordinary
poplars, is not used in growing trees except to clear sites.
However, with glyphosate-resistant trees, growers could spray
plantations with glyphosate. The scientists have also engineered hybrid poplars to produce Bt toxin—in the hopes of
controlling a serious leaf-eating pest, the cottonwood leaf
beetle. The Cooperative so far has not applied for commercial permits for the engineered trees.

A major concern of mine, which I have not yet seen discussed
in depth by genetic engineers, is the evolution of new
pathogens and diseases. This could happen when genetic engineering changes an animal's physiology and biochemistry,
which could affect harmless and beneficial, so-called symbiotic,
bacteria and other microorganisms living in that animal. They
and their metabolic products could become extremely harmful

to transgenic and normal animals. Also when a virus particle is used to carry desired genetic material into an animal or plant, there is the possibility of recombination with existing viruses in the transgenic organism to create a new virus and possibly a lethal disease. For example, the Rous sarcoma virus, which causes cancer in poultry, has been engineered to carry growth hormone genes into fish to make them grow faster. The consequences to consumers of such fish are unknown.

Fish engineer Dr. Frank Sin advises:

> The use of viral DNA sequences in constructing fusion genes to use in fish for human consumption is unwise owing to the lack of knowledge on the possible side effects of using viral DNA sequences in genetically engineered food sources. Thus, the use of "all fish" gene constructs should be mandatory if the genetically modified fish are for human consumption.[15]

In addition to the potential risks of these transgene-carrying viral vectors is the risk of using "marker" genes as tags to identify transgenic organisms. Some of the markers used are genes conveying resistance to antibiotics such as neomycin and ampicillin. The fear is that these genes could be transferred to any number of intestinal bacteria in birds, mammals, and humans who eat the transgenic plants containing these now widely used markers. For this reason the Australian government has banned the import of transgenic corn with an ampicillin-resistant bacterial gene ID tag.

The global potential for widespread genetic pollution is very real, considering the fact that countries like China have virtually no regulations governing or oversight of the release of GMOs (genetically modified organisms) into the environment. Hundreds of thousands of acres of transgenic plants are being grown in China already.[16] Though genetic engineers tend to be ignorant of environmental risk ("Let's see what happens since no one can predict . . ."), their knowledge base on gene function needs to be challenged, since it is full of holes. A major hole is their lack of understanding of what they call "junk" DNA, some 97 percent of which in human cells does not code for proteins and appears to consist of meaningless sequences. Subjecting this non-coding DNA to linguistic tests, however, reveals striking similarities to ordinary language, which means junk sequences may play important roles in the molecular biochemical language of life yet to be deciphered.[17]

The arguments of the biotechnocrats are unsound. They insist that genetic engineering is no cause for alarm because it has been occurring in nature spontaneously since the Beginning of creation and has been extensively employed by humanity in selective breeding of crops and animals. But there's a world of difference between those processes and current genetic engineering. The natural direction of plant and animal genetic development is toward ever increasing refinement, specialization, and diversity. Thus, natural development is open ended; there is no end to the Beginning.

With genetic engineering technology, there is an end. GE technology is directed away from natural creation, toward monocultures and reduced biodiversity and toward entropy rather than entelechy. Moreover, nature has erected strong barriers to cross-breeding, preventing most species (even if quite similar) from exchanging DNA. All spontaneous biological development, as well as all traditional selective breeding, operates within these boundaries. In contrast, genetic engineering shatters these species boundaries and through microsurgical implantation or other forceful procedures transposes genes between widely distinct species.

GENETICALLY ENGINEERED INSECTS

Some biotechnologists are developing techniques to genetically engineer insects, the Mediterranean fruit fly and the dengue/ yellow fever mosquito being the first successes. There are projects and visions of making more productive and disease-resistant silk worms and honey bees; of making harmful insects sterile or able to transmit disease to other undesirable insects; or even, in the process of feeding on plants and sucking the blood of humans and other animals, able to deliver protective antibodies and vaccines.[18]

My response to this is that although they are small, insects account for over five-sixths of all animal life on earth. Long before I became, as a teenager, an elected Fellow of the Royal Entomological Society in London, insects taught me that if you harm them, they will harm you.

In spite of informed opposition, in 1997 the U.S. government officially permitted the first field trial of a transgenic insect ever released into the environment, a western predatory spider mite. In the laboratory these mites mysteriously "lost" the foreign genes they had carried for over 150 generations when they were put into the field. Healthy insect populations have a resonance with co-evolved and co-dependent plants, birds, and other wildlife. Those who make some plants and insects transgenic are surely reopening Pandora's box.

People who are reluctant to acknowledge the threat of genetic engineering may be in denial, afraid to consider where the agribiotechnology industry is going and what it means in terms of biocultural diversity and the integrity and future of earth's creation. The life and beauty of the planet has intrinsic value, and for most humans, but not all, wild nature is the primary source of divine revelation and spiritual, not just material, sustenance and delight.

The new world order of GATT and the World Trade Organization serves the ideology of genetic imperialism, driven by the arrogance and greed of a multinational biotechnocracy that has already engaged in genetic piracy, as described earlier, violating the sovereignty of nations over their natural resources and the privacy rights of genetically unique tribal peoples.[19]

The biotechnology industry has lured sufficient venture capital that it can persuade governments and citizens to believe that genetic engineering is the only way, the final techno-fix, to feed the hungry world and cure a plethora of diseases. Individuals prone to breast cancer and other gene-linked diseases will soon be screened and may experience all manner of socioeconomic discrimination as a result. Yet most of these diseases are part of the circle of profit, by which an industry that poisons the populace with industrial wastes and agrochemicals benefits a subsidiary that treats those who are most genetically susceptible to carcinogenic, teratogenic, mutagenic, and xeno-chemicals (i.e., chemicals that mimic estrogens, neurotransmitters, etc.).

CHAPTER 8

Scientific and Bioethical Issues in Genetic-Engineering Biotechnology

> Every creature has its own reason to be.
> All its parts have a direct effect on one
> another, a relationship to one another,
> thereby constantly renewing the circle
> of life.
> —Johann Wolfgang von Goethe

Through genetic-engineering technology, we now have the power to profoundly alter all life forms and the very nature of nature—the natural world, or earth's creation. What are the short- and long-term consequences for humanity, animals, and nature, and what are the ethical principles and boundaries? What risks are justified by what benefits?

This new technology is complex, with many risks, costs, and benefits that need careful consideration because it could permanently and irreversibly alter the biology of life forms, the ecology, and natural evolution.

Through various techniques, the genetic composition of animals, plants, and microorganisms can be altered in ways radically different from those achieved by traditional selective breeding. Genes can be deleted, duplicated, and switched among species. Animal and human genes have been incorporated or "spliced" into the genetic structure or germ plasm of other animals, plants, bacteria, and other microorganisms. Human genes are now present in the genetic makeup of some mice, sheep, pigs, cattle, fish, and other animals.

The creation of transgenic plants, animals, and microorganisms, along with a host of other developments in genetic-engineering biotechnology, are touted as progressive, if not necessary, and as promising great benefits to society (and investors). Although I have found no coherent argument based on reason, science, or ethics to support any of these claims unconditionally, the biotechnology life-science industry and its supporters, just like the supporters of factory farming and vivisection, give enthusiastic and unconditional endorsement to new developments in biogenetic manipulation and to the industrialization and patent protection of its processes and products. The hyperbole employed on behalf of such new developments, coupled with a highly competitive and volatile world market, is driven by risk-taking venture capitalists whose cavalier attitude toward such significant risks as socioeconomic inequity, ecological damage, and animal suffering is neither progressive nor visionary. Unfortunately, this attitude is understandably often shared and rarely challenged by bioengineering scientists and academics in their employ, and by politicians and policy makers, who are generally scientifically illiterate.

This is not a good foundation for any new technology, least of all for such a profound and complex one as bioengineering. It is incumbent upon all who do not feel so sanguine about the directions this new technology is taking to challenge its assumptions and presumptions.[1] The doublethink and newspeak logic* of the biotechnocracy evidences some disturbing warning signs, notably of historical amnesia, ecological and biological illiteracy, ethical and moral dyslexia, blind faith, and ideological rigidity.

An international bioethics council within the United Nations would be a beginning to help ensure that this technology is applied with the minimum of harm to further the good of society and the integrity and future of the planetary biosphere. Insofar as its applicability to organic agriculture, biogenetic engineering is, from a philosophical perspective, anathema. It is mechanistic, deterministic, and reductionistic, while organic agriculture is seen as emulating nature—i.e., ecologistic, dynamically indeterminate, holistic, and regenerative. There is also an inimical difference in attitude that separates these two worldviews and in the kinds of medicine, industry, and market economy they aspire to. It has to do with reverential respect for the sanctity and intrinsic value of life, which is more evident on a well-operated (and well-loved) organic farm than in a biotech laboratory or on an industrial farm.

---

*E.g.: Knowledge is Power and Science is Truth. From *1984* by George Orwell.

The ideal of value-free objectivity in the method of scientific investigation provides no ethical basis for determining the risks, costs, and benefits in the technology transfer of biotechnical discoveries from the laboratory setting to the real world. A technocratic society runs the risk of serious error in believing that the "truth" of the scientific method is an ethically objective yardstick. This belief system of *scientism*, which is like a religion in the late twentieth century, accounts for the rigid "science-based" criteria and policies that corporations and governments—the entwined limbs of the technocracy—so adamantly adhere to. Yet this yardstick is as linear as it is simplistic. A broader bioethical framework is urgently needed in order for society to transcend technological enchantment, so that the fruits of scientific research may be realized for the benefit of the entire life community of the planet (see p. 185).

Like most scientific developments, genetic-engineering biotechnology is neither intrinsically evil nor inherently good. How this new knowledge is applied is what matters. It cannot be applied, however, without consideration of bioethical principles. And it cannot be objectively evaluated in isolation from the various contexts in which it will be applied.

Two contexts of particular interest to me are agriculture and the use of animals for biomedical research and biopharmaceutical industrial purposes. I am especially concerned about applications of genetic-engineering biotechnology in agriculture, because it is being applied primarily to maintain a dysfunctional system. We have an animal-, rather than a plant-based, agriculture in the industrial world, which causes much animal suffering and isn't good for the environment, for consumers, or for the social economy of rural communities. And it is now well documented that conventional agricultural practices are ecologically unsound, inhumane, and in the long term unsustainable, even with ever more costly corrective inputs. Some of these are being developed and misapplied by agricultural biotechnologists, who endeavor to maintain and expand globally a bioindustrialized food and drug industry that must be opposed by all because it fails to meet any of the following bioethical criteria of acceptability: that it be humane, ecologically sound, socially just, equitable, and sustainable. Rather, it is a major threat to biodiversity and to the social economies of many more sustainable farming communities.

Now, via GATT, the World Trade Organization, and Codex Alimentarius, the life-science industry, with its new varieties of patented seeds and other bioengineered products and processes, is moving rapidly to a global agricultural and market monopoly.

With regard to the patenting of animals, plants, and other life forms, I believe that it is demeaning to refer to them as "intellectual property" and that there are unresolved questions of ethics and equity over the patenting of life.

The spirit of enterprise and state of mind behind genetic engineering evidences an ethical blindness to the natural integrity, purity, and sanctity of being. Otherwise, how would we ever consider inserting our own and other alien genes into other species, drastically altering their nature and future to make them more useful to us rather than fulfilling their biologically ordained ecological, evolutionary, and spiritual purposes?

The domestication of plants and animals and the transformation of their habitats and ecosystems to serve human ends have had profound consequences on their nature and on the entire natural world. But do thousands of years of domestication and ecosystem alteration provide a historically valid and ethically acceptable precedent for even more profoundly altering the intrinsic nature of other living beings through genetic engineering?

We must ask: Is it necessary? Who are the primary beneficiaries? What are the direct and indirect costs and risks? Are there safer, less invasive and enduring alternatives? Does a cultural history of exploiting life justify its continuation and intensification through genetic-engineering biotechnology?

"HARD" AND "SOFT" PATHS

There are two basic paths that this new technology can take, and I have designated them as "hard" and "soft." The hard path results in permanent physical changes that may be transmissible to subsequent generations. These changes in animals' physiology or anatomy may result in their suffering. For purely ethical, humanitarian reasons, I am opposed to all hard-path applications of genetic-engineering biotechnology of which there is no demonstrable benefit to the animals themselves.

Where such benefits can be demonstrated, as in efforts to conserve endangered species and to prevent or treat various animal diseases of genetic origin, and there are no alternative strategies to achieve the same ends, then I would accept on a case-by-case basis some hard-path applications. But those applications that design animals for purely utilitarian ends should be questioned and opposed in the absence of demonstrable animal benefit.*

Likewise, any nontherapeutic product of genetic-engineering biotechnology, such as recombinant (synthetic) bovine growth hormone (rBGH), that is used to increase animals' utility and can result in animal sickness and suffering, or increase the risk thereof, is not ethically acceptable.

The creation of transgenic plants that are resistant to herbicides and virus infections, or that produce their own insecticides, belong in the hard-path category. They do not accord with accepted standards and principles of organic and sustainable agriculture and are a potential threat to wildlife and non-harmful insects, microorganisms, and biodiversity.

Soft-path developments with this technology include the creation of new-generation vaccines, veterinary pharmaceuticals and diagnostic tests, and genetic screening to identify defective genes and those that convey disease resistance and other beneficial traits. The most promising of these soft-path developments that I would endorse are immunocontraceptives, new-generation contraceptive implants for humans and other mammals.

Soft-path genetically engineered products are acceptable, provided they are safe and effective without side-effects that could cause animal suffering; provided they cannot be transmitted to or harm nontarget species (as with modified live virus vaccines);

---

*Utilitarian ends such as to increase appetite, growth, muscle mass, leanness, fertility, or milk or egg production or to deliberately create developmental abnormalities and genetic disorders.

and provided they are not used to help prevent diseases in animals kept under stressful, inhumane conditions (as on factory farms), rather than changing the conditions that contribute to increased susceptibility to disease. A full socioeconomic and environmental impact assessment is needed prior to approving these soft-path products for animal use. For example, a new vaccine for cattle to combat trypanosomiasis (to which wild ruminants are immune) could result in an unacceptable loss of biodiversity and an ecologically harmful expansion of livestock numbers.

EPIGENETIC CONCERNS

In altering the genetic makeup of any organism via genetic engineering, we run not only the risks of genetic dysfunction, disease, and pollution, but also the considerable risk of inadvertently damaging the so-called epigenetic capacity of organisms to adapt to environmental changes and stress factors. The resonance and responsiveness of organisms to the outside world is mediated in part by genes and in part by as yet little understood physiological and biochemical processes that environmental influences can modify.

This documented biological phenomenon casts an element of doubt and uncertainty over the belief that genes alone are responsible for all that is inherited from one generation to the next. Epigenetic inheritance is a fact of life, and many more examples are likely to be identified once geneticists are better able to monitor the methylation process, which has been identified as the key whereby certain genes are controlled by external environmental influences.[2] Genetic engineering could interfere with the fine-tuning of organisms via epigenetic linkage to environmental changes, which could be the reason "bad" weather had such disastrous consequences for Monsanto's genetically engineered cotton in 1998. New diseases might also develop as genetically altered plants, with altered physiology and biochemistry, affect various microorganisms on the plants and in the soil.

GENETIC DETERMINISM

The broad range of potentially beneficial applications of genetic-engineering biotechnology in agriculture and in veterinary and human medicine are being overshadowed and undermined by an overarching narrowmindedness. This is the reductionist view that since there is a genetic basis to disease, then genetic engineering is the answer to preventing and treating various human, crop, and farm-animal health problems. And that along the way we may even discover ways to genetically engineer (and patent) life forms to enhance their usefulness and "improve" their nature, be it the stature and intelligence of our own species, the growth rates of chickens and pigs, or the herbicide and pest resistance of corn and beans. This simplistic view of genetic determinism is a potentially harmful one because even though it claims to be scientific and objective, i.e., value free, it is extremely subjective and biased since it puts so much value (and faith) in the genetic approach to improving the human condition and the disease resistance and productivity of crops and farm animals.[3]

A more interdisciplinary and holistic approach to human, animal, and crop health and disease prevention is urgently needed. Seeking purely genetic solutions is too narrow and reductionistic, and because of the uncertainty principle inherent in the genotype-environment interface, genetic determinism is unlikely to bring the benefits that its proponents and investors hope and believe are possible.

In its unsubstantiated promises to feed the hungry world, and its promises of great profits for investors, genetic-engineering technology drains human resources from funding more sustainable, eco-friendly, and socially just ways of producing food. It likewise impedes the medical sciences from breaking free of a reductionistic and mechanistic paradigm of human health that blames either nature or our genes for most of our ills. Once people blamed the gods, but as Hippocrates advised, "Physician, do no harm." Conventional medicine has yet to realize this wisdom and put it into practice.

Had the dominant Western culture based its foundation on the worldview of Pythagoras or Plato, rather than that of Aristotle, with his hierarchical, linear thinking, and not on interpreting the book of Genesis as giving man unconditional dominion over God's creation, then our powers over the atoms of matter and the genes of life would probably be applied to very different ends: wholeness and healing, rather than commodification, monopoly, and selfish exploitation.

The original meaning of *dominion* in the book of Genesis does not rest in the Latin *domino*, to rule over, but in the root Hebrew verb *yorade. Yorade* means to come down to, to have humility, compassion, and communion with all of God's creation. It is an injunction of reverential care, of humane stewardship. Hence, genetic engineering is antithetical to Judeo-Christian tradition and ethics. It also violates the precept of Islam, where it is regarded as a sin to willfully interfere with God's creation, and would be considered a blasphemy of hubris to engage in creating transgenic life forms and then to go and patent them.

Genetic engineering is anathema to Buddhists, Hindus, and Jains, since it is a direct violation of the doctrine of ahimsa, of noninterference and nonharming. It is also a fundamental biological interference with the earth's creative process of natural unfoldment and thus a disruption in the spiritual process of incarnation.

One would think that an enlightened biotechnology industry would make every effort to protect the remaining integrity and biodiversity of genetic resources of the first creation—the last of the wild. Future generations, with a more sophisticated understanding of genetic engineering, will need wild places as a source of uncontaminated genetic resources. This "biobank" must be protected now and not ransacked by the industries of timber, mining, real estate, and other business enterprises, and by the millions of poor people who are malnourished and either landless or without sustainable agriculture or way of life. I have seen them in India and Africa leaving an imprint similar to that left by the clear-cutting of old-growth forests and totally obliterated prairies that the U.S. government still permits. To this destruction by the rural poor—especially from grazing too many livestock, ploughing marginal land that erodes easily, and killing trees for firewood—we must add industrial and agro-chemical pollution in both the "first" and "third" worlds.

An important step to protect the biobank is to eliminate all possibilities of genetic pollution from transgenic crops, bacteria, insects, oysters and other mollusks, shrimp, and other genetically engineered seafoods, which will be the first foods of animal origin on the market.[4] The second step must be to label all foods to indicate whether any product or ingredient has been genetically engineered. To have this information is a consumer's right, on religious and ethical grounds, since many, regardless of assurances as to food quality and safety, would prefer not to unknowingly purchase genetically modified foods. The public has a right to be informed and a right to be able to choose natural foods if they prefer, especially since genetically altered foods violate many people's religious principles.

The third step entails international cooperation on the scale of a United Nations environmental paramilitary police force to help countries protect their wildlife preserves and bio-diversity, both aquatic and terrestrial, from further human encroachment, wholesale exploitation, and genetic piracy.

There is no way to collect all potentially useful life forms and store them in culture media, or in seed, sperm, embryo, and cell banks. Many seeds lose their vitality when stored and need to be frequently germinated and harvested, genetic changes due to local environmental influences notwithstanding. They must be protected *in situ* and *in toto*.

The late Professor René Dubos, a renowned biologist from Rockefeller University, said, "An ethical attitude to the scientific study of nature readily leads to a theology of the earth." His concerns, expressed in 1972 in his book A God Within, are extremely relevant today with the advent of genetic engineering.[5] He cautioned, "A relationship to the earth based only on its use for economic enrichment is bound to result not only in its degradation but also in the devaluation of human life. This is a perversion which, if not corrected, will become a fatal disease of technological societies." Without an "ethical attitude," beginning with a reverential respect for all life and based on internationally accepted bioethical principles and values (see page 185), this disease is very likely to be fatal to the dominant culture.

The ethics of preserving the earth's bio-integrity must serve to direct and constrain the emerging biotechnocracy. The biotechnology industry must adopt these ethics; otherwise, the costs and risks to future generations will far outweigh the short-term profits of the present.

Obedience to natural law, which is based on the bioethics of sound science and moral philosophy, must be absolute, like compassion, or else it is not at all. Through science, reason, and reverence, we learn the wisdom of obedience. Industry and commerce must conform to natural law and, like human society, do nothing to jeopardize natural biodiversity, bio-integrity, or the future of earth's creation. The first task of science and of biotechnology is to begin the healing of humanity, which is biologically, economically, and spiritually dependent on the protection and restoration of what is left of the natural world: first creation first!

The application of bioethics, which is the foundation of natural law, to establishing the necessary limits and boundaries of new technologies like genetic engineering is long overdue. Every nation-state needs to have a bioethics council that would function to maximize the benefits and minimize the risks and costs of all new technologies and related commercial activities, and to ensure international harmonization of their policies and guidelines with all countries via the United Nations Council on Sustainable Development.

BEYOND GENETIC DETERMINISM AND REDUCTIONISM

Genes "intelligently" organize structural proteins into myriad environmentally co-evolved, living forms. These life forms are variously self-healing, self-replicating, even marginally self-conscious to varying degrees; and they form mutually enhancing or symbiotic communities. Collectively, for example, they help create and maintain the soil and the atmosphere that sustains the body-earth and life community, much like our digestive, circulatory, and respiratory systems are cellular communities that sustain the body-human. We find phenomenological parallels between the ecological roles of a living forest or a watershed of streams and swamps, and the functions of our own lungs, circulatory system, and kidneys.

In order to know, therefore, *how* genes, organs, and forests function, we must understand their purpose within the larger functional systems in which they participate. Therefore, we must seek to understand the *contexts* in which genes operate,

their history (or evolution and development), and their consequences. Such knowledge of temporal and spatial relationships
within the intersecting biofields of organisms and their environments is lacking in the reductionistic paradigm of conventional scientific inquiry, and in conventional medical practice,
which has been so reticent to recognize the myriad connections
between healthy forests and a healthy people. Hence, most of
our agricultural, medical, and technological inventions and
interventions have caused more harm than good.[6]

The direction being taken by the life-science industrial
biotechnocracy today, especially its investment in creating and
patenting transgenic life forms that have been engineered to
serve narrow human ends, is cause for concern, as the science
base is unsound and there is no ethical or ecological framework.
The biofields of developing organisms, what Rupert Sheldrake
calls morphic fields,[7] and the environmental influences on gene
expression are excluded from the narrow paradigm of genetic
determinism.

It is unlikely that genetically engineered crops will ever help
compensate for nutrient-deficient soils, polluted water, or a
contaminated food chain. Using biotechnology to make farm
animals more productive and efficient in the context of intensive industrial agriculture will only extend the animals' suffering and prolong the adverse environmental, economic, and
consumer health consequences of this kind of agriculture.

Genetic-engineering reductionists might find it advan-
tageous to further reduce life conceptually to its next level—

primordial energy, vital force, or chi—and then reflect upon the possibility that the final frontier of materialistic and mechanistic science, molecular genetics, is a grand illusion, a mirage created by a defective worldview and a misconception of human purpose and significance. The antidote is a paradigm shift that broadens our understanding of life by fostering a sense of reverence and awe and a feeling for the spirit or essence of life that is omnipresent in all matter and manifest in all sentient beings.

ECOLOGICAL AND SOCIAL CONCERNS

In relation to ecological concerns, I would concur with Mario Giampietro that:

> Current research on agricultural applications of genetic engineering seems to be heading exactly in the same direction as the green revolution. The main goal is to provide yet another short-term remedy to sustain, if not increase, the scale of human activity. . . . Genetic engineering aimed only at increasing economic return and technological efficiency is likely to further lower the compatibility of human activity and natural ecosystem processes. . . . Before introducing a massive flow of new transgenic organisms into the biosphere, a better understanding of the endangered equilibrium of the biosphere should be achieved.[8]

Philosopher, scientist, and activist Vandana Shiva eloquently expresses my concerns over the harmful consequences of this new technology and the need for public input to minimize potential harm:

My major concern these days is with the protection of cultural and biological diversity. I am preoccupied with the ecological and social impacts of globalization of the economy through free trade on the one hand and the colonization of life through genetic engineering and patents on life forms on the other hand. My sense is that unless we can put limits and boundaries on commercial activity and on new technologies, the violence against nature and against people will become uncontrollable. The question I constantly ask myself is, What are the creative catalytic linkages that strengthen community and enable communities of people to exercise social and ecological control of economic and technological processes?[9]

One of the major risks of genetic-engineering biotechnology has a conceptual basis that Craig Holdrege thoroughly dissects in his book *Genetics and the Manipulation of Life*.[10] It stems from scientific reductionism, objectivism, and the mechanomorphizing and reification of genetic and developmental processes and shows no concern and responsibility for effects on the organism and the environment. The belief in genetic determinism is as dangerous ethically as it is flawed scientifically because it is based on the central dogma that genes alone determine how an organism develops and functions.[11] The antidote that Holdrege offers is in seeking an understanding of relationships via contextual thinking, based in part on regarding heredity as potential or plasticity complemented by heredity as limitation or specificity. *Genetics and the Manipulation of Life* is an important book for all students of the biological sciences and for those proponents and critics of biotechnology in particular.

We must be mindful of the fact that nothing that exists originated independently. Therefore, all existences are ultimately interconnected, co-evolved, and interdependent. Genes are not the sole or even the primary controllers and regulators of life processes. It is a product of hubris and reductionism that in isolating and manipulating DNA, we believe we can gain control over life. If we do not act quickly to address all the factors that are leading to the death of nature, then the virtual reality that the global life-science industrial biotechnocracy is fabricating will collapse. We have neither the wisdom nor the resources to develop a viable analog of the earth's atmosphere, or of an old-growth forest, a mountain stream, or a coral reef. The Arizona desert–based Biosphere II Project could not even create an earth-in-microcosm with a stable atmosphere using mainly natural, living components from the real world, and with the best minds and technology that money could buy. Biosphere II ran out of oxygen.

How then can we expect unnatural, genetically engineered life forms to do any better in the virtual world of global industrialization, even when we too are engineered to withstand the harmful, somatic effects of chemicals, pathogens, and radiation?

SOME BIOETHICAL CONCERNS AND SOLUTIONS

I am deeply concerned by what I see as a lack of vision in the agricultural biotechnology industry, which is limiting its benefits to humanity and its potential for profitability and sustainability. The cavalier attitude of corporations, governments, and much of academia toward the release and commercialization of transgenic crops is especially troubling. A related concern is over the fact that agricultural biotechnology is focused primarily on major commodity crops and not linked in any significant way with ecologically sound and sustainable crop and livestock husbandry. It therefore cannot play any significant role in helping relieve world hunger or, especially, in implementing appropriate practices and inputs to restore agricultural and rangelands now sorely degraded worldwide.

Lester Brown writes in *State of the World 1994* that University of Minnesota agricultural economist Vernon Ruttan summarized the feeling of a forum of the world's leading agricultural scientists when he said, "Advances in conventional technology will remain the primary source of growth in crop and animal production over the next quarter century." Biotechnology should not be seen as a panacea, or as a substitute for conventional technologies, the most basic of which are good farming practices in accordance with the land ethic and the principles of humane sustainable agriculture. My opposition to conventional agricultural biotechnology is based on its evident band-aid and high-input roles in conventional, nonsustainable agriculture. As such, it represents a major obstacle to the research, development, and adoption of more sustainable, ecologically sound, and in the long-term more profitable farming practices.

As for advances in human and veterinary medicine, safer and more effective new vaccines, pharmaceuticals, and diagnostic tools are some of the positive benefits of biotechnology. As discussed earlier, however, the creation (and patenting) of transgenic mice to serve as models for human diseases is to be questioned, since only 2 percent of human diseases are caused by a single gene defect. A similar reductionistic approach, which reveals a naive genetic determinism,[12] is evident in the human genome project and similar projects on farm animals aimed at identifying "good" and "bad" genes.

The conservative Hastings Center has published a report that details the complexity of bioethics, especially the creation of genetically engineered animals.[13] This report emphasizes the difficulties of developing a "grand monistic scheme" that "establishes a hierarchy of values and obligations under the hegemony of one ultimate value." Such an approach to dealing with contemporary ethical concerns is dismissed by the authors because, while it "may serve the peace of the soul by reducing internal moral conflict," it would, they believe, work only in relatively small and homogeneous communities. It "invariably is bought at the price of the variety and richness of human experience and significant cultural activity. In this sense it impoverishes the human soul."

I would argue to the contrary. There are moral absolutes such as reverence for life, compassion, and ahimsa (nonharmfulness) that can provide both a goal and a common ground for a reasoned and scientific approach to resolving ethical issues. These absolutes are the cornerstones of a monistic hierarchy of human values that could effectively incorporate the plurality of interests of various segments of society and of different cultures.

The Hastings report suggests an alternative strategy: "a coordination of values both *within* and *among* spheres of activity."

It goes on to say, "Contextually coordinating our plural obligations requires a decision-making art of moral ecology, judicious weighing of the several obligations in the various contexts at hand be they narrower or wider."

But what catalyst, what shared value or concern is to bring people with differing points of view or value systems together, and against what template is the "judicious weighing" to be done? Surely without the shared goal of enhancing the life and beauty of the earth, based on the holistic principles of bioethics that extend concern for the good of society to the good of the planet, and that give fair consideration to both human and nonhuman (plant and animal) communities, "the decision-making art of moral ecology" will accomplish little. Such a shared goal as enhancing the life and beauty of the earth is based upon the supreme ethic of reverence for all life, as well as ahimsa. Once such a shared goal is realized and the moral ecology of diverse value systems democratically integrated via mutual understanding and respect, then a "grand monistic scheme" that establishes a hierarchy of values and obligations under the hegemony of one ultimate value—reverence for all life—would be possible, if not inevitable.

To aim for less is to fall short of realizing the full political and spiritual power of human reason and compassion that the discipline of bioethics embodies. By focusing primarily on the "moral ecology" of diverse and often opposing human interests and values, a human-centered or cosmopolitan, rather than an eco-centered or cosmocentric, worldview and template for public policy will emerge; and will suffer the gridlock of opposing and unreconciled values in the absence of a monistic unifying principle, such as ahimsa.

In spite of these limitations in not envisioning the ultimate integration of diverse moral values and human interests into a "grand monistic scheme," the Hastings report takes a significant step in that direction, noting that:

> We require systematic ethical responses that genuinely recognize the plural value and ethical dimensions of our worldly existence. How do we square this circle, which is demanded by our overall responsibilities to humans, animals, and nature? How should such practical decisions be substantively guided? This is an outstanding and unsettled issue. Yet we may begin to see our way. The first clues come from the sheer plurality of practices, contexts, values, and obligations themselves. . . . We must become ethically committed, as an overarching and fundamental moral duty, to this plurality itself: to upholding and promoting the various abiding and culturally significant spheres of human activity amidst the ecosystemic life and animate world in which they are embedded. . . . Ethical atomism or provincialism is practically impossible and ethically irresponsible. Rather we must *concurrently* pursue the human, animal, and natural good. First and foremost we must prevent the significant undermining of any one domain or sphere of activity, human or natural, for the sake of others. This involves a mutual commitment, sensitivity, and concern among different human actors with various contextually defined allegiances. Such coordination requires a mutual accommodation

without forgoing fundamental value and ethical commitments. We must fashion an ethically and publicly responsible life that is broadly "cosmopolitan."

Such a human-centered or "cosmopolitan" framework for bioethics is a transitional democratization of ethical values that incorporates the moral ecology or plurality of interests and obligations of contemporary society. It fails, however, to offer any transformative principle (such as ahimsa) in unconditionally "upholding and promoting the various abiding and culturally significant spheres of human activity," many of which do violence to the sanctity of life in the name of cultural tradition and social progress.

The ultimate values and benefit of GET may well be to force us to address a host of ethical questions and to facilitate our maturation from an ego-centered or anthropocentric worldview to one that is more eco-centric and creation-centered. We ask, from this latter perspective, How can we use this and other new technologies to alleviate human and animal suffering, to heal the earth and ourselves in the process? From this "anthropocosmic" worldview, we extend the Golden Rule to all living beings. And we put into law and practice the ancient covenant of good husbandry: When we take care of the animals and the land, the animals and the land will take care of us.

THE COHERENCE OF SCIENCE, MIND, AND NATURE

The human mind—and spirit, too—is inextricably linked in its biological evolution with the natural world. To be deprived during the developmental years of contact with nature in any

meaningful, participatory way, must surely have adverse consequences on our mental development, on the refinement of our
perception, language, imagination, and creativity. The complexity of nature nurtures and enriches our personalities and
culture, enabling us to conceptualize holistically, to think ecologically in terms of complex interrelationships, processes,
causes, and consequences. For many, still, wild nature provides
the most immediate connection with divinity, a sense of the
sacred. Nature is also the matrix, therefore, of the "natural
mind" and feral vision of *Homo sapiens*, and we suffer the consequences when we destroy this matrix or grow up without significant immersion in it.

To *cohere* means to be connected naturally or logically.
Unlike other animals, our coherence, our connectedness with
nature, depends on our careful, deliberated, and respectful par-
ticipation in the broader biotic or life community. Thus when
we engage in scientific investigations of life processes and seek
to commercialize various discoveries, that knowledge—as
power over life and creation—should be applied with care and
respect for the integrity and sanctity of the biotic community.

I wonder, when people talk about the new "information
age," precisely how this information has been derived, what its
source or sources are, and to what ends it will be applied. For
instance, the genetic information derived from the reductionistic science of genetic-engineering technology represents one
extremely narrow, if not distorted, dialog with other living organisms. It is derivative of a mechanistic and dominionistic world-

view that tries to fit the complex organic process of natural food production into a more simplistic industrial process of production. There can be no meaningful dialog with nature when the worldview or paradigm—e.g., industrial agriculture—is not coherent, connected naturally or logically, with the biological reality of complex organic systems and processes.[14] What information we have about the nature of these systems and processes is first obtained through scientific investigations based on the industrial paradigm, in which increasing productivity is the primary goal. Such goal-directed research is neither objective nor impartial, and thus it is not strictly scientific or in accord with the scientific method of investigation. Basic scientific research into natural systems and processes establishes more of a dialog and is one antidote to the chaotic consequences of an incoherent and instrumental relationship with nature.

The fundamental flaw of goal-directed research and development is that no real dialog is desired, no deep understanding is sought beyond manipulation and control for human ends. Thus, motives and values are the subjective components of applied industrial and biomedical science that need to be balanced by bioethical considerations and constraints and by the insights of impartial, basic scientific inquiry. By establishing such balance, goal-directed research and development, be it in agriculture, medicine, or other human endeavor, will not be subject to the law of unforeseen consequences, so often harmful to the life community, and the best intentions of good people will not continue to go all awry.

Nature's complexity is the antidote to the simplistic, reductionistic, and mechanistic thinking that is becoming so widespread today, especially in the applications of genetic-engineering biotechnology in medicine and agriculture. The preservation of the natural world, therefore, stems from enlightened self-interest if we are not to lose our natural minds in the process of creating a virtual reality. That reality, which is taking shape today, will consist primarily of a bio-industrialized world populated by genetically engineered plants and animals designed to satisfy the insatiable needs of virtual people.

CONCLUSIONS

There is a plurality of human spheres of interest, values, and concerns, from industry and academia to the religious and the secular. The advent of GET is creating a new resonance or biofeedback between humans and the rest of earth's creation that may facilitate our moral development and ethical evolution, as it forces us to consider our duties and responsibilities toward each other, animals, and nature. Out of the plurality and diversity of human values and goals, wants and needs, GET may be the catalyst for helping us develop a more unifying life- or creation-centered bioethics. This will help ensure that the benefits of GET may be realized and in the process help humanity achieve a quality of life that is sustainable and not at the expense of animals and nature. GET runs the risk of further distancing us from animals and nature at a time when we urgently need to deepen our ethical allegiance with the natural world. Through GET we could create a new virtual reality—virtual nature, virtual animals, and virtual food—while in the process the natural world is transformed into a bio-industrialized wasteland. Is it wise to use limited public resources to fund this technology and to facilitate the expansion of industrial agriculture, aquaculture, bioremediation, and conventional allopathic medicine, while neglecting our collective, corporate, and public responsibilities to help save the oceans; clean up the environment; protect biodiversity; restore natural ecosystems; and promote organic, humane, and sustainable agriculture?

As Nicanor Perlas states, "The biotechnology revolution is upon us. It promises to be the new economic infrastructure of the post-industrial era. We have to enter into the labyrinths of its hidden agenda, grab the beast by the horns, and transform it to truly serve the interests of nature, society, and the human spirit."[15] Perlas argues convincingly that we must add a spiritual dimension to the social, environmental, and animal-welfare considerations in our assessment and approval of biotechnology, in terms of the origin and future of creation and the meaning, values, and purpose of human existence.

The good of society and the good of the earth, the future of humanity and the future of creation, are interdependent and not mutually exclusive. A universalizing bioethics that includes such principles as ahimsa, compassion, reverential respect for all life, social justice, eco-justice, sustainability, protection of biodiversity and cultural diversity, and enhancement of the life and beauty of the earth will blossom if there is openness, honesty, and trust in evaluating and approving GET and other technologies that are developed to help improve the human condition and the state of the world.

> Where is the wisdom we have lost
> in knowledge? Where is the knowledge
> we have lost in information?
> —T. S. Eliot

> Error has turned beasts into men.
> Will truth be able to turn men back
> into beasts?
> —Friedrich Wilhelm Nietzsche

In October 1996 the Alliance for Bio-Integrity was incorporated, and it is achieving significant progress in coordinating scientists, public-interest organizations, and religious groups to take a stand for stronger regulation of genetic engineering. This campaign will have a significant practical effect in the form of a lawsuit against the FDA to obtain mandatory labeling and stricter safety testing of genetically engineered foods. The executive director of the alliance is Steven Druker. Visit the alliance on the web at www.bio-integrity.org; phone it at (800) 549-2131; or write it at P.O. Box 2927, Iowa City, IA 52244-2927.

BIOETHICAL EVALUATION AND PRINCIPLES

Various developments and applications of genetic-engineering technology (GET) should be objectively and rigorously evaluated and opposed when the following bioethical criteria for acceptability are not fully met:

1. NECESSITY: Is the new technology, product, or service really necessary, safe, and effective, and are there alternatives of lesser risk and cost?
2. TRACKING: Can released genetically engineered life forms be identified, traced, contained, and recalled if needed?
3. OVERSIGHT AND COMPLIANCE: Can the new technology, product, or service be effectively regulated to maximize benefits and minimize risks, and at what cost to society?
4. PUBLIC DEMAND AND ACCEPTANCE: Cultural, religious, ethical, health, and safety concerns must be fully considered and respected.
5. ENVIRONMENTAL IMPACT: Short- and long-term consequences and influence on wild plant and animal (including invertebrate) species and microorganisms must be rigorously evaluated.
6. ECONOMIC IMPACT AND VIABILITY, SOCIAL JUSTICE, AND EQUITY (INTERNATIONAL AND INTERGENERATIONAL): Who will benefit? Who might be harmed?
7. SOCIAL AND CULTURAL CONSEQUENCES: What is the impact on the structure of agriculture, nationally and internationally, and on more sustainable traditional and alternative agricultural practices at home and abroad? Is it technologically appropriate and respectful of cultural pluralism, and does it enhance the development of human capacity and potential?
8. ANIMAL WELFARE: Will the new product or service enhance animal health and overall well-being?

The ethos (intrinsic nature) and telos (natural role and purpose) of every life form are at risk of being engineered to meet the pecuniary ends of industrial society. The ecos, or natural world, to which animals, plants, and microorganisms belong and which they help maintain, is at risk of being obliterated in the process. But when we expand our simplistic assessment of biotechnology from risks, costs, and benefits—from a purely economic point of corporate self-interest—and incorporate bioethical principles and criteria, the necessary paradigm shift or change in worldview will occur. This is essential if the benefits of this new technology are to be fully realized to serve the interests of nature, society, and the human spirit, precisely because biotechnology is based on the reductionist principles of the inorganic, physical, and mechanical sciences that focus on manipulating, controlling, and directing complex life processes to meet purely human ends.

CHAPTER 9

Corporate Ethics in Biotechnology: Hard Choices, Soft Paths

> The commercialisation of science in genetic engineering biotechnology has compromised the integrity of scientists, reduced organisms including human beings to commodities . . . It results in a monolithic wasteland of genetic determinist mentality that is the beginning of the brave new world.
> —Mae-Wan Ho

C orporations such as Monsanto and DuPont do not stand alone in believing that the ultimate good for humanity is in sustainable economic development. While critics contend that this is an oxymoron, the corporate and government consensus that emerged from the Rio Earth Summit in 1992 is very clear: sustainable development is the top priority for both the developing and the industrialized worlds.

While the definition of sustainable agriculture is being hotly debated, be it low- or high-input, ecological, biodynamic, organic, etc., the fact remains that a conceptual seed has been planted—namely, that of sustainability. Before this seeding, which could mean the opening of the corporate mind, the core values of agribusiness and of the corporate-industrial ethos were economic growth and ever greater productivity and efficiency, with monopoly and patent protection of intellectual-property rights providing the necessary security. Without stockholder security and funds from private and public sectors (government tax revenues) to stimulate research and development and the creation of new products, services, and market niches, the spirit of enterprise would be crushed.

U.S.-based multinational life-science corporations enjoy the security of government and even military protection to help ensure that America will stay competitive in the global marketplace and have a quality of life that is the envy of the rest of the world. Through NAFTA and GATT, the influence and interests of multinational corporations will transcend those of nation-states, and a new world order, based, we hope, on "planetary patriotism," is in the making.

Still embryonic, the corporate seed of sustainability must be nurtured, and in the realm of agricultural and industrial biotechnology, the global "greening" of multinationals and of lending institutions such as the World Bank and International Monetary Fund can become a reality. Enlightened corporate self-interest is linked with sustainable development, just as free and fair trade agreements hold the promise of international peace and cooperation rather than competition, conflict, and war. The failure of the petrochemical-based Green Revolution, because it was neither sustainable nor socially just and did not primarily benefit small farms and village communities, is a lesson that corporations should not ignore.

Critics of the "industrial semioticians" who see the changing image of corporations and international banks as merely Orwellian "newspeak" and "doublethink" rhetoric, while business goes on as usual, need to look behind this window dressing. They will find a few naked emperors and some with new clothes, but by and large, when the imperial arrogance of the biotechnocrats is peeled away, and the hyped promises of medical and agricultural miracles through biotechnology scrutinized, the limited, profit-driven horizons of the industrial establishment are clear for all to see. Like any living organism, the corporation, regardless of its cultural roots and transnational connections, is as vulnerable as it is opportunistic. Survival and growth (or multiplication) are based on the ecological (and economic) principles of sustainability. No organism is secure if it exhausts or poisons its habitat or resource base. The scolex-headed worldview of technological determinism and its opportunistic, pioneer ethos is an evolutionary dead end.* The devolved condition of the evolutionary path that some corporations have taken contrasts with the alternative path of increasing complexity and diversity, biological principles that enlightened corporations with enhanced computer technology and information systems, including satellite monitoring, are evolving to embrace. The diversification of the petrochemical,

---

*The scolex is the eyeless and brainless head of a tape worm, often equipped with hooks and suckers.

pharmaceutical, and food-industrial complex into a life-science industry is a natural evolutionary progression following decades of publicly and privately funded research in genetics and molecular biology. Safer alternatives to conventional products and processes that are not based on nonrenewable resources such as petroleum are some of the promises of biotechnology.

Safety and effectiveness are not the sole criteria for government approval and public acceptability of new biotechnology products, however. But as Greg Simon, chief domestic policy advisor for Vice President Al Gore, has opined, "I predict that if Europeans insist on blocking a safe product"—such as recombinant bovine growth hormone—"for social and economic reasons, they'll see a flight of capital in biotechnology like they'll never believe." This posture by the Clinton administration, aimed at protecting and promoting U.S.-based biotechnology multinationals, is a continuation of the Bush administration's White House Council on Competitiveness, chaired by Dan Quayle. U.S. biotech companies were thrown into disarray when President Bush was asked to sign on to the international treaty to protect biodiversity. The fear was that U.S. companies would have limited access to the plant and animal species that would be protected as intellectual property under that treaty. That fear eventually dissipated because, as Carl Feldbaum, president of the Biotechnology Industry Organization, observed, "We support the treaty, because biodiversity is the lifeblood of biotechnology." President Clinton signed the biodiversity treaty in June 1993. Yet even so, the United States is still not a member-country endorsing this treaty because the Senate has blocked final ratification.

These turnarounds are noteworthy since, as I emphasized in my book *Superpigs and Wondercorn*, a government that is too protective of domestic interests is likely to do more harm than good to those interests in relation to the global marketplace and the volatile nature of venture capitalism.[1]

The indecision over accepting the ethical and economic ramifications of the biological-diversity treaty is indicative of the internal conflicts of interest and of vision with which the biotechnology industry, and its cadre of lawyers, scientists, and chief executives, had been wrestling. The U.S. government led a minority of countries to sabotage the United Nation's international Biosafety Protocol Treaty at a meeting in Colombia in February 1999. Along with Australia, Argentina, Canada, Uruguay, and Chile, the United States blocked what 130 nations wanted to include in the treaty: the right of every country to know which seeds, grains, and foods that it imports have been genetically engineered.

The corporate mind in America has yet to evolve and realize that it is enlightened corporate self-interest to explore new avenues of global cooperation. The old paradigm of competition is an anachronism to the evolving new world order, and the U.S. biotechnology industry has created a Jurassic Park with its outmoded thinking, based on a worldview of economic determinism and monopolistic control of the earth's genetic resources.

The gene rush is a latter-day gold rush, but the corporate cornucopia of biotechnology products and profits will not be forthcoming even with intellectual property rights protection if there is no social justice and public trust. This is the first lesson that this new industry has learned: The genetic resources of nation-states cannot be pirated, and governments must, therefore, become more responsive to the rights and interests of other countries and indigenous peoples.

The bioethical complexity of genetic-engineering biotechnology is demanding on many levels, from the regulatory oversight and compliance levels to the social and economic levels. As such, it not only challenges the corporate mind in a manner unprecedented by technological innovation, but also is a potent stimulus for the development of the corporate mind toward a worldview that is not as narrowly goal-oriented or directed by short-term profit margins and risk-benefit ratios, but instead considers broader bioethical criteria, as outlined in the previous chapter.

The biofeedback that this new biotechnology creates, especially for its advocates and adopters, mandates the development of ecological awareness. Theodore Roszak writes in *Voice of the Earth*, "The core of the mind is the ecological unconscious. For ecopsychology, repression of the ecological unconscious is the deepest root of collusive madness in industrial society; open access to the ecological unconscious is the path to sanity."[2] For Roszak, this path leads us to realize the synergistic interplay between planetary and personal well-being and to accept that "the needs of the planet are the needs of the person, the rights of the person are the rights of the planet."

In his seminal book *The Dream of the Earth*, Thomas Berry concludes that human and earth technologies must be integrated.[3] This is the essence of enlightened corporate behavior and self-interest. In order to accomplish this challenging task, the corporate mind, its ethos and telos, must be changed.

The seed concept of sustainability that agricultural biotechnocrats are now broadcasting will find fertile soil, not in the public's acceptance or government support of genetic-engineering biotechnology, but in the life-science industry itself creating the optimal environment to maximize the benefits and profits of this new technology for all. This optimal environment is best assured by subjecting every proposed new biotechnology product to rigorous evaluation on the basis of the bioethical criteria listed on p. 185. With this approach, a soft path of enlightened corporate self-interest will be laid, as distinct from a hard path whose products, such as synthetic bovine growth hormone and herbicide-resistant seeds, are created solely for their developers and investors. There has been little resistance to genetically engineered insulin and "vegetable rennet" because those products follow the soft path.

After the unforeseen farmer and consumer backlash that Monsanto faced in bringing rBGH* to the farmer's gate, the pitfalls of the hard path are clearly evident. Government approval of rBGH does not mean that the floodgates have been opened for new GE products and patented varieties of transgenic crops, farm animals, and microorganisms. On the contrary, both start-up biotech companies and big multinational corporations that have invested heavily in genetic engineering will have to follow the soft path. The rBGH experience is a clear

---

*Monsanto's trademark name for its genetically engineered bovine growth hormone is Posilac.

signal to the industry to do its homework before investing in the R&D for a product that will do violence to any of the bioethical criteria for acceptability. Corporations that do their homework well will realize the goal of sustainability.

Through genetic-engineering biotechnology we have God-like powers, and also choices and responsibilities. Players in the corporate world, in meeting the challenge of making wise and responsible choices, will indeed become planetary patriots. Evidence of humility, compassion, and ethical sensibility may be lacking behind the veneer of hyped superpigs, wondercorn, and other touted miracles of bioindustrial technology. But as biosophy* teaches us that the needs of the planet are the needs of the people, so too the concept of sustainability depends on the realization that the interests of the corporation are the interests of people and planet alike. Considerable wealth, security, and fulfillment can be had by directing this new technology to serve the greater good, to help heal the earth and humanity. The alternative hard path will only quicken the social, ethical, economic, environmental, and physical collapse of our institutions and life-support systems.

As I concluded in *Superpigs and Wondercorn*, "Although 'environmentally neutral' biotechnology is a valid goal, a better one would be an environmentally and socially enhancing biotechnology. This is an attainable goal. It is not wishful thinking. Mistakes will be made, but such risks can be minimized and this new technology can be applied creatively and profitably if

---

Biosophy refers to the combined wisdom of the biological sciences.

there is corporate responsibility. And this responsibility for the integrity and future of creation is indeed as great as the power we now have over the gene and over life itself."

Can we say that society today is really functional? Leaders speak of the need for more family values, jobs, health care, and education, but all to what end if everything that matters is in limbo in the moral vacuum of economism and consumerism? Such a vacuum trivializes the human condition and purpose. The despair, alienation, violence, escapism, and suicides of our youth today are clearly products and symptoms of our dysfunctional condition. This condition will continue to deteriorate until ethics and morality are woven back into the fabric of the professions and the corporate world—and into the World Bank, the White House, and the WTO and GATT. A laissez-faire, market-driven global economy where growth is both a means and an end in itself will be the nemesis and the apotheosis of industrial civilization. What will remain of our sanity, of our humanity, and of the life and beauty of the natural world depends on what we can protect and cherish today, under the broad banners of bioethics and biocultural diversity, through international accord. This accord can come only when governments and corporations become transparent by working openly and in concert with the people, who have a right to be fully and fairly represented in all policy decisions that affect the life community.

Recent developments in agricultural biotechnology illustrate clearly how the decision-making process of corporations and government is framed within an outmoded anthropocentric paradigm that is purportedly "science-based" but precludes such bioethical considerations as socioeconomic, ecological, and moral consequences.[4] A moratorium on the approval of genetically engineered bovine growth hormone for use in dairy cows and on the patenting of genetically engineered animals and plants was put in place by the European Parliament almost a decade ago precisely because of such bioethical concerns, while the U.S. government rushed to approve both.

Current risk-benefit analyses and scientific studies to determine the safety and effectiveness of new biotechnology products and processes, such as potatoes and corn that produce their own pesticides, virus-resistant squash, herbicide-resistant soybeans, porcine growth hormone, and transgenic salmon and catfish, are too simplistic. Without a paradigm shift that makes the science base broader and more relevant, and includes bioethics in it, the real benefits of biotechnology may never be fully realized.

THE MONOPOLY GAME

As a result of recent corporate mergers and acquisitions, the food and drug industry is being consolidated into a small consortium of global multinational monopolies. Of the emerging biotech giants, Pioneer Hi-Bred International, with operations in sixty-five countries, is the largest of the big three, being linked with DuPont Chemical Co. Second is Monsanto, which has bought up several seed and biotech companies and sought to merge with one of the world's largest pharmaceutical and home-care products, American Home Products, to create a $96-billion life-science conglomerate. But this merger fell through in 1998.

Number three is Swiss-based Novartis, which is the product of a merger between two drug and chemical giants, Sandoz and Ciba-Geigy, and is aggressively promoting genetically engineered crops with the other two giants worldwide.

Six chemical companies now dominate research and development in plant genomics (genetic analysis and engineering): Monsanto, DuPont, Sandoz, Zeneca, Ciba-Geigy, and Enimont. These companies, along with Shell, W.R. Grace, and Cargill (the world's largest grain and oil seed trader), now dominate the international seed market. In June 1998, Monsanto bought Cargill's international seed operations, in Central and South America, Europe, Asia, and Africa, for $1.4 billion.

The control of the world's food and health by a small monopoly of big MACCs (multinational agricultural chemical companies) will become a reality if governments do not acknowledge the serious economic, social, and environmental consequences of corporate hegemony and support the rights and interests of an informed citizenry. As Bill Heffernan, professor of rural sociology at the University of Columbia, Missouri, sees it, "Before long they'll tell the farmer: 'We won't buy your product unless you use that [our] seed.' A few big farms will thrive, but most of the small ones will disappear" (*Christian Science Monitor*, July 30, 1998).

GOVERNMENT INDUSTRY ALLIANCE

The life-science industry, and particularly its agribusiness sector, does not know where to draw the line because its scope is unconstrained by ethical considerations and government oversight and regulation. The U.S. industry has achieved almost total government deregulation as well as assistance in developing and selling its technology and products to developing countries and in avoiding the costs of being regulated and accountable, because that would weaken their position in the competitive world marketplace.

The Edmonds Institute* and the Third World Network posted the following information to show how closely the U.S. government and the Clinton administration, like the prior Bush administration, is linked with the agricultural biotechnology industry:

DAVID W. BEIER—former head of Government Affairs for Genentech, Inc., now chief domestic policy advisor to Al Gore, Vice President of the United States.

LINDA J. FISHER—former Assistant Administrator of the United States Environmental Protection Agency's Office of Pollution Prevention, Pesticides, and Toxic Substances, now Vice President of Government and Public Affairs for Monsanto Corporation.

L. VAL GIDINGS—former biotechnology regulator and (biosafety) negotiator at the United States Department of Agriculture (USDA/APHIS), now Vice President for Food & Agriculture of the Biotechnology Industry Organization (BIO).

MARCIA HALE—former assistant to the President of the United States and director for intergovernmental affairs, now Director of International Government Affairs for Monsanto Corporation.

MICHAEL (MICKEY) KANTOR—former Secretary of the United

*Located at 20319 92nd Avenue West, Edmonds, WA 98020.

States Department of Commerce and former Trade Representative of the United States, now member of the board of directors of Monsanto Corporation.

JOSH KING—former director of production for White House events, now director of global communications in the Washington, DC, office of Monsanto Corporation.

TERRY MEDLEY—former administrator of the Animal and Plant Health Inspection Service (APHIS) of the United States Department of Agriculture, former chair and vice-chair of the United States Department of Agriculture Biotechnology Council, former member of the U.S. Food and Drug Administration (FDA) food advisory committee, and now Director of Regulatory and External Affairs of DuPont Corporation's Agricultural Enterprise.

MARGARET MILLER—former chemical laboratory supervisor for Monsanto, now Deputy Director of Human Food Safety and Consultative Services, New Animal Drug Evaluation Office, Center for Veterinary Medicine in the U.S. Food and Drug Administration.*

WILLIAM D. RUCKELSHAUS—former chief administrator of the U.S. Environmental Protection Agency (USEPA), now (and for the past 12 years) a member of the board of directors of Monsanto Corporation.

MICHAEL TAYLOR—former legal advisor to the U.S. Food and

---

*Margaret Miller and Michael Taylor (and Suzanne Sechen, an FDA "primary reviewer for all rBST and other dairy drug production applications") were the subjects of a U.S. General Accounting Office investigation in 1994 for their role in FDA's approval of Posilac, Monsanto's formulation of recombinant bovine growth hormone. The GAO found "no conflicting financial interests with respect to the drug's approval" and only "one minor deviation from now superseded FDA regulations." (Quotations are from the 1994 GAO report.)

Drug Administration's Bureau of Medical Devices and Bureau of Foods, later executive assistant to the Commissioner of the FDA, still later a partner at the law firm of King & Spaulding where he supervised a nine-lawyer group whose clients included Monsanto Agricultural Company, still later Deputy Commissioner for Policy at the U.S. Food and Drug Administration, and now again with the law firm of King & Spaulding.

LIDIA WATRUD—former microbial biotechnology researcher at Monsanto Corporation in St. Louis, Missouri, now with the U.S. Environmental Protection Agency Environmental Effects Laboratory, Western Ecology Division.

CLAYTON K. YEUTTER—former Secretary of the U.S. Department of Agriculture, former U.S. Trade Representative (who led the U.S. team in negotiating the U.S. Canada Free Trade Agreement and helped launch the Uruguay Round of the GATT negotiations), now a member of the board of directors of Mycogen Corporation, whose majority owner is Dow AgroSciences, a wholly owned subsidiary of the Dow Chemical Company.

CHAPTER 10

Where Do We Draw the Line?

> Totalitarian ideologies, which degrade
> Man by reducing him to an object
> while breaking human rights, raises in
> a worrisome way certain uses of the
> potentials offered by biotechnology.
>
> —Pope John Paul II

U ntil the later part of the twentieth century, little thought has been given to how we ought to treat animals and other living beings, besides humans. If there had been such consideration, mother pigs would be able to root and wallow freely and play with their piglets. But hundreds of millions of farm animals around the world now cannot even walk or turn around because they are kept in such extreme confinement on factory farms for their entire lives. Such conditions, seen and judged through the narrow lens of animal production science, are normative and, according to contextual or situational ethics, are neither wrong nor cruel because they're predicated upon the assumption that animals don't have feelings, a conscious self, or, therefore, any personhood.

Many people, including those who still eat pigs and other animals, feel such treatment is cruel and unnecessary confinement. Many believe that animals have an intrinsic nature, a soul, that they can and do suffer, and that they therefore have a natural right to be treated with respect and compassion. Each being's ethos, or intrinsic nature, is part of the earthly and universal manifestation of the creative spirit, the *ethos dei*, according to my interpretation of creation-centered spirituality. It is only from this kind of paradigm or worldview that a firm bioethical bridge of empathy and understanding can be created between humans and the rest of creation to prevent further harm to both, to foster reverence for the sanctity of life, and to stop the desecration of what remains of wild nature.

Because genetic-engineering biotechnology is being applied primarily within the domain of a worldview that is antithetical to this spirituality, and is devoid of any bioethical sensibility and self-constraint, I am deeply concerned. I hope this book awakens such concern more widely, because with the right bioethical template, genetic engineering could be used in many beneficial ways: to create transgenic bananas and spinach that when eaten protect people from hepatitis, rabies, and other epidemic diseases; to engineer new and safer alternatives to conventional methods of human and nonhuman birth control; and to develop new DNA-based diagnostic tests for various acquired and inherited diseases, for example. These kinds of applications of genetic engineering and biotechnology are ethically acceptable insofar as they contribute directly to helping improve human and animal health and well-being.

I draw the line—actually, a circle—around such compassionate applications of new technologies. Outside the circle, and where I hope society will also draw the line, are virtually all the other current and foreseeable applications of biotechnology described in this book. These include the engineering of humans, other animals, and also plants in the realm of "genetic enhancement," for example, to make humans taller, stronger, and more intelligent; pigs and other farm animals fatten faster and be ever more productive; and crops more nutritious, containing touted "nutriceuticals." Why are these things outside the circle? Because genes alone do not determine the better qualities of humanity, and human eugenics is fundamentally flawed by the genetic determinists' limited, reductionistic view of what it takes to become, and what it means to be, human. Because, for ecological and consumer health reasons, we should all drastically reduce, if not eliminate, animal fats and proteins from our diets; and, therefore, we should not try to make the species we eat ever more productive through genetic enhancement. As for genetic enhancement of crops, their nutrient value is determined primarily by soil quality, which is currently declining as a consequence of industrial, chemical-based methods of food production that are now incorporating GEOs, the antithesis of organic, ecological farming methods.

Using genetic engineering to increase disease resistance in crops (including trees) and food animals (including aquatic species) is ethically unacceptable when the ways in which the crops and farm animals are raised and their genetic makeup (a result of selective breeding for desired production traits, such as rapid growth and early maturity) decrease their disease and pest resistance and result in much animal stress and suffering. The state of mind that rationalizes animal suffering is, I believe, a worse disease than any of the human diseases animal experimentation seeks to cure, and is itself a cause of much human illness and suffering.

We must also question the use of genetic engineering to permanently alter defective germ-lines in humans and all other species. Regardless of altruistic motives, the high attendant risks of such *genetic remediation* (compared to the potential benefits of less invasive treatments such as bone marrow and other somatic cell and alternative delivery systems of genetic material that preclude direct interference with the germ-line) also put us on the slippery slope of eugenics.

Engineering bacteria and other microorganisms to convert toxic industrial wastes into less toxic materials, a process called *bioremediation,* is not ethically acceptable if it means the continuation and expansion of industrial activities that produce harmful products and byproducts.

Engineering mice and other species to serve as analog models for various human diseases in order to develop new drugs is ethically and medically questionable when those diseases could be prevented by addressing anthropogenic causes such as chemical pollution, excess saturated fats and animal proteins in the diet, and other unhealthy consumer habits, addictions, and life styles.

Creating transgenic "pharm" animals and cloning them to mass-produce new health-care products in their milk, serum, and urine is ethically questionable. It is nothing less than *genetic parasitism* when we put human genes into other animals for medical and other commercial purposes. Nor are there any guarantees that these animals will not suffer developmental and physiological abnormalities and even succumb to new diseases as a consequence of their genetic constitution being altered. A more humane alternative is to create transgenic plants and bacteria for such purposes, but with the proviso that adequate safeguards are put in place to prevent accidental release into the environment and consequential genetic pollution.

I also draw the line at using biotechnology to maintain endangered species in captivity, as by embryo transfer, cloning, and transgenics, when there is little or no effort being taken to protect these animals in the wild; when there would be no place natural for such captive species ever to be released; and when the public is led to believe and support such activities as real conservation rather than seeing the artificial perpetuation of "virtual" species in captivity as a tragic symbol of our harmful impact on the natural world and the animal kingdom. Cloning endangered species and resurrecting already extinct ones from naturally preserved DNA is not conservation but a grand illusion of god-play, and of vested human interest in preserving *in vivo* unique genetic material that may have some future commercial utility.

What of rich people cloning their pets, like the couple who gave $2.3 million to genetic engineers at Texas A&M University in 1998 to find a way to clone their beloved old mixed-breed dog? How much more good that money could do supporting an animal shelter or refuge such as the one my wife operates in a poor rural community in southern India! And what of all the good dogs waiting to find a home and the millions that are euthanized every year?

On the basis of the ethical and ecological concerns expressed in this book and elsewhere, a moratorium on the release of any and all transgenic life forms into the environment and a total recall of all that have been released so far are warranted. The precautionary principle has been totally ignored by the life-science industry, which is putting the entire natural world at risk from transgenic bacteria and other microorganisms, plants, insects, mollusks, fish, and other GEOs that have been deliberately released simply to make money. This judgment is not too harsh, considering the fact that this industry is engaging in genetic piracy, essentially depriving third world farmers of any intellectual property rights to their own indigenous varieties of seeds and various natural products that they have developed over centuries themselves; and of any equity in this stolen "intellectual property," which various multinational corporations have acquired and gained legal protection of via worldwide patents and devious manipulation of international trade and marketing agreements.

There is probably a loss in genetic plasticity and potential in GE plants when transgenic changes such as Bt-toxin production and herbicide resistance reduce drought and cold resistance. The reported lower fertility and fecundity of some widely planted GE crops today by Marc Lappé and Britt Bailey[1] do not bode well for future food security.

There is clearly no vestige of social justice, altruism, or intent to help feed the hungry world because the name of the game is corporate growth and profits through monopoly. I despised the game Monopoly as a child when I saw how some of my peers behaved as they were getting rich and others were getting poor. The adult version of the game is much worse—serious and potentially deadly to those on the fringe or in the trenches, and to corporate and government whistle-blowers.

In addition to engaging in genetic piracy, industrial agri-biotechnology's multinational oligopolies such as Novartis, DuPont, and Monsanto are spending billions of dollars not to help feed the starving world but to lobby governments, support political candidates, establish academic chairs and university departments, and brainwash the public, through the media they manipulate, that it's all for the good: A better world through biotechnology. In science we must trust.

"COUNTRYCIDE"

There is absolutely no question that raising genetically engineered (transgenic) crops will cause more harm to wildlife, especially to insects, small mammals, and birds, than conventionally and organically raised crops. First, consider the fact that of the almost 70 millon estimated acres of genetically engineered (GE) crops planted in 1998 around the world, an estimated 71 percent are herbicide resistant. This means all fields planted with herbicide-resistant soybean (that is 52 percent of the total of GE crops),[2] canola, and corn will be sprayed with toxic weed killers. In 1998 some 51.3 million acres of GE crops were grown in the United States.* All wild plants, and the insects, birds, and other creatures that depend upon them—the last vestige of biodiversity in our rural "countryside"—will be wiped out. This is "countrycide." In developing countries such as India, where Monsanto is taking root, weeds—i.e., wild plants—are also an important source of fodder for poor villagers' and tribals' livestock, which are employed as weeders. Furthermore, these herbicides are harmful to aquatic life and may harm the human immune and neuroendocrine systems.

Insect-resistant (Bt) corn (planted on 24 percent of the current acreage of almost 70 million) and insect (Bt) and herbicide-resistant cotton will cause further harm to wildlife because these transgenic crops produce their own pesticide (Bt). Even more serious harm to the ecology of the soil may result, because this Bt toxin does not rapidly degrade in the soil after the crop has been harvested and the remains used as compost. Recent research has found that this toxin in transgenic crops is likely to accumulate more and more in the soil after each crop, and will poison many beneficial insects and other organisms essential

---

*Dr. Susan Harlander, vice president of Green Giant/Progresso agriculture research and development, estimates that in the year 2000, 50 percent of all field crops in the United States will be genetically engineered (*Feedstuffs*, April 26, 1999, p.8).

for the ecological health and fertility of the soil. This could effect the nutritive value of crops grown in sterilized soil.[3]

Agribiotechnology, as it is currently being applied and forced onto the third world through trade deals, great promises, and bribes (and even enjoying White House and U.S. intelligence support), is ethically unacceptable and contemptible. Creating an empire out of greed and insensitivity, in total disregard for the rights of aboriginal peoples, is leading to the loss of biological diversity and cultural diversity at an accelerating rate. All to what end? The death of nature, the end of natural evolution.

CHAPTER 11

Saving the Seeds
of Humanity from
Spiritual Corruption

> You are a God only insofar as you
> recognize yourself to be a human being.
>
> —Plutarch

> Three kinds of progress are significant
> for culture: progress in knowledge and
> technology; progress in the socialization
> of man; progress in spirituality.
> The last is the most important.
>
> —Albert Schweitzer

We are crossing the boundaries that separate us from other species, not by way of empathy but through genetic engineering. Instead of contemplation, we engage in manipulation. Where there was once communion and wisdom, there is now control and information. Life, once held sacred, is now a patentable commodity. Biotechnology, applied with reverence and humility, may do some good, but lacking empathy and ethical sensibility, and serving as a means to gratify pecuniary ends and insatiable wants, it can only do more harm.

Animals, trees, certain rock formations, and other natural creations had such great intrinsic value and empowering qualities that our ancestors were moved to awe and reverence. Where is the awe and reverence in the genetic engineer's laboratory, where human beings play God, putting their own genes into other creatures? Where is the awe and reverence in the bioconcentration camps and fields of corn and cows, pigs and pines? Where are the voices that cry "obscene" and call for an end to keeping animals so confined that they cannot even walk or turn around? What of the silence of the birds in these agroforests and over the plains and prairies now almost all gone under the plow? What will be left soon to sustain the human spirit? The appetites that corrupt our souls do no less to the soul of the earth.

The fruits of the tree of life will be ever fewer and more bitter, and the false fruits of the virtual reality of the emerging biotechnocracy—genetically engineered and irradiated foods— will not sustain us in either body or spirit.

An imperialistic biotechnocracy can never see why respecting animals' rights and having a sacramentalist attitude toward nature and all of God's earthly creation are keys to our own health and the realization of our intrinsic divinity.

I feel a growing sense of urgency over the closely linked spiritual, social, and environmental crises of these times because the corruption of individuals, institutions, public-interest organizations, industry, commerce, politics, and governments worldwide is intensifying. I see corruption being dismissed as greed and rationalized as the means of doing business to achieve justifiable ends.

Having worked in one developing country (India) and also being involved in animal and environmental protection in the United States, where—on both hemispheres—corruption is either denied or fatalistically accepted as normal, I am more aware than ever of the consequences of confronting and exposing corruption in whatever form it may take.

Fear, as well as arrogance and greed, corrupt the spirit. Fearing losing influential political allies and corporate sponsors, leaders of public-interest organizations, from animal and environmental protection to consumer health and worker safety (especially when they get big and fear getting sued), become corrupted. They rationalize their "neutrality" and diminishing advocacy by claiming "scientific" objectivity and "political correctness."

So I appeal to those who dare to be politically incorrect, who see ethical consistency as a virtue, and who still find in nature, animals and plants, their sacraments and a heart for communion, to stand up and be recognized—we are not alone!

Artist, philosopher, and physician Frederick Frank put it this way:

> Reverence for Life implies the insight, the empathy and compassion that mark the maturation of the human inner process and that implies overcoming the split between thinking and feeling that is the bane of our scientism and the idolization of

technology that distances—estranges—us from all emotional
and ethical constraints. This same distancing, this objectifica-
tion of the unobjectifiable, is characteristic of all Realpolitik,
racism, ethnic cleansing, cruelty and exploitation of the other
by political, racial, religious collectivized in-group egos, including that free-market mentality for which all that is, is looked
upon as mere raw material-for-profit, even if it ruins our
species and our earth for generations to come.[1]

As a veterinarian with over thirty-five years' experience, I can
vouch for the decline in compassion and concern for animals
and nature by government, industry, and commerce, while public concern for animal welfare and environmental protection
has clearly increased. Medical and scientific experts have been
fired by the legalized-mafia of their governments and corporations for raising questions about pesticides, dioxins, and various
genetically engineered products. A veterinarian in Japan, who
voiced concern about the Yakuzas' (Japanese mafia's) refusal to
allow their fighting dogs to be given a local anesthetic before
their wounds were sown up, was recently killed. He was killed
for questioning the authority of power. A Belgian veterinarian
investigating the illegal use of drugs by the veal industry was
killed in 1996 by the European drug mafia. Chico Mendez,
leader of sustainable forest gatherers, was killed by cattle ranch-
ers less than a decade ago for opposing the clearing of the Amazon rain forest.

The corruption of the human spirit—of our humanity, our
dignity, and our integrity—by a powerful few who seek control
over the economy or over our religious and political beliefs, and

thus over our lives, must be fought on all fronts lest we lose all that makes us human and gives life meaning. This battle has confronted every human civilization since the beginnings of agriculture and recorded history.

How well educated we are and from what culture and socioeconomic class we come make little difference for those of us who share the kinship of having enjoyed the affection of animals and the intimacies and mysteries of wild nature. I have sat with Indian jungle tribals and rejoiced with them over saving a valuable and beloved old milk cow from a difficult labor; stood beside an Eskimo whale hunter in communion with the Arctic summer night; and played in the boundless circle musicians make, beating drums and blowing flutes and didjeridoos with indigenous peoples from around the world to celebrate the life and beauty of the earth. The spirit of humanity is not yet dead. But I wonder what will become of us as a species when the materialistic monoculture of consumerism spans the globe.

Many children already have no real closeness with any animals, wild or tame, except in picture books and in the virtual realities of zoos and videos. Even the parts and products of the animals they consume seem to have no connection with laying hens, milk cows and fat pigs, chickens, lambs, and young cattle, or with the living soil. Would it be a reality check—or psychologically harmful—for them to see just how these animals are raised today in factory farms and feedlots? And to see just how they are transported for slaughter and killed?

A few fortunate children have animals as companions, not just as "pets" or toys. Their parents love and value animals in and for themselves and see the animals as part of the family and deserving of respect and equal consideration. Few children ever see wild nature, and they have decreasing contact with the natural world.

Few parents teach their children reverence for all life, opening their hearts to the wonders and mysteries of wild nature. Few children now go out to hunt and trap and fish with their fathers. Few fathers dare to show their loving concern for an injured fawn or hunting dog—or for a child confused by the sudden realization that we must all kill to live. It can be difficult to empathize with those who never learned why they must kill a deer swiftly with one arrow, and not just for sport; and with those people who still eat other animals without a second thought. But empathize we must to help restore our *collective humanity.*

Those awakened by the intimacies of a participatory relationship with the natural world—with wild nature and all our animal and plant relations, wild and tame—recover the spirit of their humanity. The more our hearts open to the sacramental powers of the natural world, the more we must be prepared to suffer the pains and sorrows of a sentient world that is subjected to the cruel and selfish dominion of our fallen species, which Frederick Frank sees as being afflicted by the spiritual virus of contempt for life. Only then, I believe, can we be effective in confronting industrial agriculture's biotechnology and multinational corporate hegemony and have the vision and passion to implement alternatives such as organic agriculture and bioregionally autonomous stewardship and protection of cultural and natural resources.

It is not mere nostalgia for the cry of the loon and the howl of the wolf that calls us to preserve the wilderness. It is a deeper, atavistic wisdom. When our minds are no longer stopped or our hearts filled by sunsets and loving dogs, by flying geese and waterfalls, will we still be human? What would we care about then? What will become of us in the world we are now creating, where there will be no wild nature left on a human-infested, plundered, and poisoned planet?

What is to become of us we are witness to now: The more disconnected from nature and the land we become, the more violence and destruction we see, and the more our youth feel disconnected, alienated, unfocused, and uninspired by the shallow market-driven values of a consumptive culture.

One man I know, who as an adolescent had his first epiphany in a redwood forest, now leads a North American Indian coalition to save the last of the U.S. Northwest's old-growth rain forests. These ancient, natural forests around the world could soon be logged and seeded with new transgenic supertrees, all of one kind, but without natural forests there might be no corn in Kansas, the White House could be under water, and populations could be devastated by global warming's new diseases.

Save the forest for our own sakes? Or for the forest's sake? Either way is right as long as we all unite against the dark side of human nature that can rationalize the obliteration of the old-growth forests, the depletion and pollution of the oceans, and the end of the wild; sanction the wholesale exploitation and suffering of animals; permit the subjugation and annihilation of indigenous peoples and cultures; and even sanctify the bio-engineering of new life to serve the needs and wants of its depraved appetites.

How indeed will future generations judge these times? What values will they derive from the kind of world and culture they inherit, or will they too, like most of us, their forebears, choose to live in ignorance and denial?

Enjoying the sacraments of creation and communion with nature are not ends in themselves; otherwise, they amount to nothing more than self-indulgence and idolatry. Communion with the God of nature and with the nature of God as manifest on earth and in all sentient beings, moves us to live and to fight for justice for all creatures and for creation; to help save the last of the wild; and to alleviate and prevent the suffering of other sentient beings. This fight for justice for all that is sacred is the fight for the truth of *equalitarianism*. Since all of life is sacred, all living beings should be given equal consideration. This calls for radical compassion, nonviolent action, and spiritual anarchy in our personal lives and in our communities. We see it as a quickening and "greening" grassroots movement today. It goes beyond the polemics of animal rights versus human interests, and nature-conservation versus economic growth and material affluence, because the fight for truth redefines what it means to be human.

To be human means to be a part of the whole, part holy, part humus. We cease to be well and to be human when we wantonly destroy the whole and when our chauvinism defiles all that is holy, including our own humanity. So to be human means to realize the divinity of nature and self, and to be mindful of the God in all, as we are all in God.

With great genetic expectations, the life-science industry is playing Monopoly with Mother Earth and all of God's creation. At least they think they are. In actuality, they are playing Russian roulette. Without a radical change in consciousness, as the poet T. S. Eliot warned, we will go out "not with a bang but a whimper." Former UN secretary general U Thant saw the whimper as a "planet running out of air, food, and pure water."

We know what is happening, and I do not believe that we need be fatalistic. Nor are we helpless. We must not become callous or indifferent, live in denial, or even rationalize one protective and destructive belief system after another, such as the importance of industrial and economic growth and other forms of insanity including racism, speciesism, and dominionism.

We can stop the "bang" through global conciliation, and address the "whimper" with immediate planetary CPR: conservation, preservation, and restoration of the natural world.

Genetic engineering and other technologies could play a role in saving the earth via CPR. But they will not and cannot until we utilize such instrumental and empirical knowledge within a framework of bioethics. Several basic bioethical principles have been identified in this book.[2] A spiritual awakening and redefinition of what it means to be human are also imperative.[3]

When we reaffirm the sacramental value and powers of plants, animals, and the rest of God's creation the seeds of our humanity may be saved. With the passionate light of radical compassion in action that is our reaffirmation to help prevent animal suffering and the loss of natural biodiversity, these seeds will be saved, along with all of nature, which connects our spirits with the manifest divinity of a living earth.

Notes

INTRODUCTION

[1]See: M. W. Fox, *Eating with Conscience: The Bioethics of Food*, Troutdale, OR: NewSage Press, 1997.

CHAPTER 1

[1]For further discussion see: M. W. Fox, *The Boundless Circle: Caring for Creatures and Creation*, Wheaton, IL: Quest Books, 1996.

[2]Benjamin Farrington, *Francis Bacon: Philosopher of Industrial Science*, New York: Henry Schuman, 1949.

[3]See: Van Rensselaer Potter, "Global bioethics: Linking genes to ethical behavior," *Perspectives in Biology and Medicine* 39 (1995): 118–131; see also M. W. Fox, *Global Bioethics: Interdependence, Evolution, Compassion* (book in progress).

[4]For further discussion see: Bernard E. Rollin, *The Unheeded Cry: Animal Consciousness, Animal Pain, and Science*, Oxford, England: Oxford University Press, 1989.

[5]S. L. Davis and P. R. Cheake, "Do domestic animals have minds and the ability to think? A provisional sample of opinions and questions," *Journal of Animal Science* 76 (1998): 2072–79.

[6]A. Kimbrell, *The Human Body Shop*, New York: Harper Collins, 1993.

[7]Available from Public Employees for Environmental Responsibility, 810 First Street NE, Suite 680, Washington, D.C. 20002.

[8]*Biotech Reporter*, February 1996.

CHAPTER 2

[1]D. J. Grant, "Ag. now driven by biotechnology," *Illinois Agri-News*, September 11, 1998.

[2]J. Klindt et al., "Growth, Body Composition, and Endocrine Responses to Chronic Administration of Insulin-Like Growth Factor 1 and (or) Porcine Growth Hormone in Pigs," *Journal of Animal Science* 76 (1998):2368–81.

[3]N. Myers, *Perverse Subsidies: Tax Dollars Undercutting Our Economies and Environments Alike*, Winnipeg, Manitoba: International Institute for Sustainable Development, 1998.

[4]K. Dawkins, *Gene Wars: The Politics of Biotechnology*, New York: Seven Stories Press, 1997; see also V. Shiva, *Biopiracy: The Plunder of Nature and Knowledge*, Boston: South End Press, 1997.

[5]M. W. Fox, *Superpigs and Wondercorn: The Brave New World of Biotechnology and Where It All May Lead*, New York: Lyons & Burford, 1992; see also D. Suzuki and P. Knudtson, *Genethics: The Clash Between the New Genetic and Human Values*, Cambridge, MA: Harvard University Press, 1990; and J. Doyle, *Altered Harvest*, New York: Viking, 1985.

[6]M. W. Fox, *Eating with Conscience: The Bioethics of Food*, Troutdale, OR: NewSage Press, 1997.

[7]N. Myers, *Perverse Subsidies: Tax Dollars Undercutting Our Economies and Environments Alike*, Winnipeg, Manitoba: International Institute for Sustainable Development, 1998.

[8]N. D. Barnard, A. Nicholson, and J. Lil Howard, "The medical costs attributable to meat consumption," *Preventive Medicine* 24 (1995): 646–55.

⁹See: N. J. Temple and D. P. Burkitt, *Western Diseaes: Their Dietary Prevention*, Totowa, NJ: Humana Press, 1994; see also J. Chesworth (ed.), *The Ecology of Health: Identifying Issues and Alternatives*, Thousand Oaks, CA: Sage Publications, 1996.

CHAPTER 3

¹Neal D. Barnard et al., "The medical costs attributable to meat consumption," *Preventive Medicine* 24 (1995):1–10.

²According to a 1986 report by the Office of Technology Assessment, in *Feedstuffs*, March 14, 1994.

³Dennis T. Avery, *Biodiversity: Saving Species with Biotechnology*, Indianapolis, IN: Hudson Institute, 1993.

⁴T. I. Hewitt and K. R. Smith, *Intensive Agriculture and Environmental Quality: Examining the Newest Agricultural Myth*, Greenbelt, MD: Henry A. Wallace Institute for Alternative Agriculture, 1955.

⁵Web site: http://vm.cfsan.fda.gov/~lrd/biocon.html

CHAPTER 4

¹V. Shiva, *Monocultures of the Mind: Perspectives on Biodiversity and Biotechnology*, London: Zed Books, 1993.

²See: T. P. Lyons and K. A. Jacques, Proceedings of Alltech's tenth annual symposium, *Biotechnology in the Feed Industry*, Nottingham, England: Nottingham University Press, 1994; see also National Academy of Science Symposium, *Metabolic Modifiers: Effects on the Nutrient Requirements of Food-Producing Animals*, Washington, DC: National Academy Press, 1994.

CHAPTER 5

¹V. G. Pursel, "Progress in genetic modification of farm animals," chapter 18 in: K. H. Engel et al. (eds.), *Genetically Modified Foods: Safety Issues*, Symposium Series 605, American Chemical Society, 1995.

[2]C. Patience et al., "Infection of human cells by an endogenous retrovirus of pigs," *Nature Medicine* 3(1997):282–86.

[3]P. R. Wills, "Transgenic animals and prion diseases," *New Zealand Veterinary Journal* 43(1995):86–87; and B. D. O'Neil, ibid, p. 88.

[4]"Transgenic chicken technology," *Genetic Engineering News*, July 1998, pp. 17–39.

[5]"Gene mutation provides more meat on the hoof," *Science* 277(1997):1922–33.

[6]*New Scientist*, October 22, 1994, p.4.

[7]For further details see: M. W. Fox, *Eating with Conscience: The Bioethics of Food*, Troutdale, OR: NewSage Press, 1997.

[8]T. D. Etherton, "Growth hormone technology develops new twist," *Nature Biotechnology* 15(1997):1248.

[9]S. Watson, "Fox in the cow barn," *The Nation*, June 8, 1998, p. 20.

[10]Environmental Research Foundation, PO Box 5036, Annapolis, MD 21403. email: erf@rachel.org

[11]K.H.S. Campbell et al., "Sheep cloned by nuclear transfer from a cultured cell line," *Nature* 380(1996):64–66.

[12]R. D. Palmiter et al., "Metallothionein-Human GH Fusion Genes Stimulate Growth of Mice," *Science* 222(1983):809–14; see also J. Berlanga et al., *Acta Biotechnologica* 13(1993):361–71; G. Brem and R. Wanke in A. C. Beynen and H. A. Solleveld (eds.), *New Developments in Biosciences: Their Implications for Laboratory Animal Science*, Dordrecht, The Netherlands: Martinus Nijhoff, 1988, pp. 93-98.

[13]"First Dolly, now headless tadpoles," *Science* 278(1997):798.

[14]See: E. Marshall, "Gene therapy's growing pains," *Science* 264(1995):1050–55; and E. Marshall, "Less hype, more biology needed for gene therapy," *Science* 270(1995):1751.

[15]See: P. B. Thompson, *Food Biotechnology in Ethical Perspective*, London: Chapman Hall, 1997; B. Rollin, *The Frankenstein Syndrome*, Cambridge: Cambridge University Press, 1995; A. Kimbrell, *The*

Human Body Shop, New York: Harper Collins, 1993; R. Hubbard and E. Walde, *Exploding the Gene Myth*, Boston: Beacon Press, 1993; and S. Krimsky, *Biotechnics and Society: The Rise of Industrial Genetics*, New York: Praeger, 1991.

CHAPTER 6

[1] *Cancer Weekly Plus*, via News Edge Corp., April 8, 1998.

[2] M. W. Fox, *Eating with Conscience: The Bioethics of Food*, Troutdale, OR: NewSage Press, 1997.

[3] C. McKee et al., "Production of biologically active salmon calcitonin in the milk of transgenic rabbits," *Nature Biotechnology* 16(1998):647–49.

[4] P. B. Thompson, *Food Biotechnology in Ethical Perspective*, London, England: Chapman Hall, 1997.

[5] R. Goldburg, "Something Fishy," *Gene Exchange* (Union of Concerned Scientists), Summer 1998, p.6.

[6] Genetic engineering news email: rwolfson@concentric.net (November 14, 1998).

[7] T. J. Hoban and P. A. Kendall, USDA Extension Service, *Consumer Attitudes About the Use of Biotechnology in Agriculture and Food Production*, Washington, D.C., July 1992.

[8] *Eurobarometer Survey*, London, 46.1.

[9] See: M. W. Fox, *Eating with Conscience: The Bioethics of Food*, Troutdale, OR: NewSage Press, 1997.

[10] V. Shiva, *Biopiracy: The Plunder of Nature and Knowledge*, Boston, MA: South End Press, 1997.

[11] *New Scientist*, February 14, 1998, pp. 14–15.

[12] R. Jefferson, "Apomixis: A social revolution for Agriculture?" *Biotechnology and Development Monitor*, no. 19 (1994) pp. 14–16.

[13] S. Nec and R. May, "Extinction and the loss of evolutionary history," *Science* 278(1997):692–94.

[14]*The Guardian*, February 16, 1999, p. 8.

[15]*The Guardian*, February 12, 1999, p. 6.

[16]Ho, Mae-Wan, *Genetic Engineering Dream or Nightmare?* Bath, England: Gateway Books, 1998. See also: Allison R. C. Thompson, and P. Ahlquist, "Regeneration of a functional RNA genome by recombination between deletion mutants and requirement for cow-pea chlorotic mottle virus 3a and coat genes for systemic infection," *Proceedings National Academy of Sciences* 87(1990):1820–24.

[17]T. Traavik, *Too early may be too late: Ecological risks associated with the use of naked DNA as a biological tool for research, production, and therapy.* Research report for DN, Directorate for Nature Management, Trondheim, Norway, 1999.

CHAPTER 7

[1]See: P. Raeburn, *The Last Harvest: The Genetic Gamble that Threatens to Destroy American Agriculture*, New York: Simon & Schuster, 1995; J. Rissler and M. Mellon, *Perils Amidst the Promise: Ecological Risks of Transgenic Crops in a Global Market*, Cambridge, MA: Union of Concerned Scientists, 1991; S. Krimsky and R. Wrubel, *Agricultural Biotechnology and the Environment*, New York: Ingram Books, 1996; and J. Doyle et al., "Effects of genetically engineered micro-organisms on microbial population and processes in natural habitats," *Advances in Applied Microbiology* 40(1995):237–41.

[2]Andre de Kathen, *Gentechnik in Entwicklungslandern: Ein Uberblick: Landwirtschaft*, Berlin, Germany: Umweltbundesamt, 1996.

[3]T. Mikkelsen et al., *Nature* 380(1996):31.

[4]See: B. Kneen, "Misguided canola update," *Rams Horn 148* (May 1997). Box 3028, Mission, British Columbia, Canada V2V 4J3.

[5]J. Rissler and M. Mellon (eds.), "Bt Cotton Fails to Control Boll-worm," *Gene Exchange* (Union of Concerned Scientists) 7, no. 1 (1996):1.

[6]*Campaign for Mandatory Labeling and Long-Term Testing of All Genetically Engineered Foods,* October 13, 1997, Natural Law Party, 500 Wilbrod Street, Ottawa, Ontario, Canada K1N GNS.

[7]*New England Journal of Medicine* 334(1996):688–92.

[8]*The Canberra Times,* June 19, 1996.

[9]R. Stone, *Science,* December 2, 1994, pp. 1472–73.

[10]A. Greene and R. Allison, *Science,* March 11, 1994, p. 1423.

[11]For a detailed review on this topic, see: A. Snow and P. M. Palma, "Commercialization of transgenic plants: Potential ecological risks," *Bioscience* 47(1997):86–96.

[12]J. Bergelson, C. B. Purrington, and G. Wichmann, "Promiscuity in transgenic plants," *Nature* 395(1998):25.

[13]M. T. Holmes and E. R. Ingham, "The effects of genetically engineered microbes on soil food webs," *Abstract: Ecological Society of America* 25, no. 2 (1994).

[14]Reported in the *Frankfurter Rundschau* by Dietmar Ostermann, December 5, 1997.

[15]F. Y. T. Sin, "Transgenic fish," *Reviews in Fish Biology and Fisheries* 7(1997):417–41.

[16]T. Plafker, "First Biotech Safety Rules Don't Deter Chinese Efforts," *Science* 266 (1994):966–67; see also A. Krattiger, "The Field Testing and Commercialization of Genetically Engineered Plants: A Review of Worldwide Data," pp. 247-66 in: A. Krattiger and A. Rosemarin (eds.), *Biosafety for Sustainable Agriculture: Sharing Biotechnology Regulatory Experiences of the Western Hemisphere,* International Service for the Acquisition of Agri-Biotech Applications, Ithaca, NY, and the Stockholm Environment Institute, Stockholm: 1994.

[17]F. Flam, "Hints of a Language in Junk DNA," *Science* 266 1994:1320.

[18]"The insects are coming," *Nature Biotechnology* 16(1998):530–33.

[19]V. Shiva, *Biopiracy: The Plunder of Nature and Knowledge,* Boston, MA: South End Press, 1997.

CHAPTER 8

[1]See: M. W. Fox, *Superpigs and Wondercorn: The Brave New World of Biotechnology and Where It All May Lead*, New York: Lyons and Burford, 1992.

[2]For more discussion see: E. Jobonka and M. Lamb, *Epigenetic Inheritance and Evolution: The Lamarkian Dimension*, New York: Oxford University Press, 1995; see also M. Pembrey, "Imprinting and transgenerational modulation of gene expression: Human growth as a model," *Acta* 45(1996):111; and M. Monk, "Epigenetic programming of differential gene expression in development and evolution," *Developmental Genetics* 17(1995):188.

[3]See: R. Leowontin, *The Doctrine of DNA*, New York: Praeger Books, 1993; see also R. Heinberg, *Cloning the Buddha: The Moral Impact of Biotechnology*, Wheaton, IL: Quest Books, 1999.

[4]For a review on this subject, see: E. M. Hallerman and A. R. Kapuscinski, "Potential impact of transgenic and genetically manipulated fish on natural populations: Addressing the uncertainties through field testing," pp. 93–112 in: J. G. Gould and G. H. Thorgaard (eds.), *Genetic Conservation of Salmonid Fishes*, New York: Plenum Press, 1993; see also A. R. Kapuscinski and E. M. Hallerman, "Implications of introduction of transgenic fish into natural ecosystems," *Canadian Journal of Fisheries and Aquatic Sciences* 48(1991):Suppl.1, pp. 99–107.

[5]R. Dubos, *A God Within*, New York: Scribners, 1972.

[6]See: J. Chesworth (ed.), *The Ecology of Health*, Thousand Oaks, CA: Sage Publications, 1996, p.19.

[7]R. Sheldrake, *A New Science of Life: The Hypothesis of Formative Causation*, London: Blond & Briggs, 1981, p.20.

[8]M. Giampietro, "Sustainability and technological development in agriculture," *BioScience* 44(1994):677–89.

[9]*Utne Reader*. January/February 1995, p. 81. See also: V. Shiva, *Biopiracy: The Plunder of Nature and Knowledge*, Boston, MA: South End Press, 1997.

[10]C. Holdrege, *Genetics and the Manipulation of Life*, Hudson, NY: Lindisfarne Press, 1996.

[11]See: R. Hubbard and E. Walde, *Exploding the Gene Myth*, Boston: Beacon Press, 1993.

[12]For an excellent discussion on this subject, see: R. Strohman, "Epigenesis: The missing beat in biotechnology?" *Bio/Technology* 12(1994):156–64.

[13]S. Donnelley, C. R. McCarthy, and R. Singleton, Jr., *The Brave New World of Biotechnology*, Raleigh-Durham, NC: Hastings Center Report, Special Supplement 24:1.

[14]See: T. Berry, *The Dream of the Earth*, San Francisco: Sierra Books, 1988; D. Abram, *The Spell of the Sensuous*, New York: Random House, 1996; and M. W. Fox, *The Boundless Circle: Caring for Creatures and Creation*, Wheaton, IL: Quest Books, 1996.

[15]N. Perlas, *Overcoming Illusions About Biotechnology*, Penang, Malaysia: Third World Press, 1994, p. 29.

CHAPTER 9

[1]M. W. Fox, *Superpigs and Wondercorn*, New York: Lyons and Burford, 1992, p. 169.

[2]T. Roszak, *Voice of the Earth*, New York: Simon & Schuster, 1992.

[3]T. Berry, *The Dream of the Earth*, San Francisco: Sierra Books, 1988.

[4]See: T. S. Kuhn, *The Structure of Scientific Revolutions*, Chicago, IL: University of Chicago Press, 1970.

CHAPTER 10

[1]Lappé, M., and B. Bailey, *Against the Grain: Biotechnology and the Corporate Takeover of Your Food*, Monroe, ME: Common Courage Press, 1998.

[2]*Gene Exchange* (Union of Concerned Scientists), Fall/Winter 1998, pp. 7 and 13.

³Crecchio, C., and G. Stotzky, "Insecticidal activity and biodegradation of the toxin from *Bacillus thuringiensis subsp. kurstaki* bound to humic acids from soil," *Soil Biology and Biochemistry* 30(1998): 463–70.

CHAPTER 11

¹F. Frank et al., *What Does It Mean To Be Human?* New York: Circumstantial Productions Publishing, 1998, p. 16.

²For further reading, see: Van R. Potter and P. J. Whitehouse, "Deep and global bioethics for a livable third millennium," *The Scientist*, January 5, 1998, p.9.

³M. W. Fox, *The Boundless Circle: Caring for Creatures and Creation*, Wheaton, IL: Quest Books, 1997.

Glossary

ANTIBODY A protein that binds to an invading antigen in the body prior to the destruction of the antigen.

ANTIGEN A foreign protein that triggers the body's production of an antibody directed specifically to neutralize the antigen.

BACTERIA Single-celled microscopic organisms with a primitive nucleus, most of which are beneficial.

BASE(S) The four structural subunits of DNA or RNA comprised of adenine, thymine, cytosine, and guanine, or in the case of RNA uracil.

BIODIVERSITY The uncounted population of earth's organisms and their genes. Refers also to the concept of the interrelatedness and interdependence of genes, organisms, communities, and ecosystems.

BIOREACTOR Term coined by industry for transgenic farm animals developed for "pharming" purposes—production of pharmaceuticals.

BT TOXIN The proteins derived from some strains of bacteria called *Bacillus thuringiensis* that become poisonous in the alkaline environment of an insect larvae's intestinal tract.

CHROMOSOMES The rod-like structures within every cell of the body carrying the genes, the number of which vary from species to species. Humans have a set of 46 chromosomes in every cell; each chromosome carries some 100,000 to 200,000 genes.

CLONE One or more genetically identical organisms. Bacteria clone naturally. Frogs and calves have been artificially cloned from somatic (body) cells.

CROSS-POLLINATION Pollen from one plant is transferred by wind or insects to the female part of a plant of a different genetic makeup.

DNA Deoxyribonucleic acid, a complex chain-like nucleotide shaped in the form of a double helix. A gene is a segment of DNA. The particular sequencing of chemicals in a segment of DNA determines what information is stored in a particular gene.

DNA SEQUENCE The linear array of bases (ATCG) that essentially spells out the genetic code.

ECOSYSTEM The composite of all the life forms and geophysical features of a given place that make up its defining qualities and type.

ENZYME A chemical that catalyzes biological reactions. Some of these chemicals, called restriction enzymes, are used like scissors to cut a desired gene segment of DNA. Other enzymes, called ligases, are used to splice the isolated gene into the DNA of a living organism, such as a bacterium. The bacterium then multiplies to produce billions of identical clones, as in the manufacture of insulin.

EPIDEMIOLOGY The study of diseases among populations.

EUGENICS The ideology of improving a species through genetic selection, manipulation, or enhancement.

GE Genetically Engineered, usually used to describe products— GE corn, for example. May also be referred to as GM (genetically modified), or GMO (genetically modified organism).

GENE FLOW Exchange of genes between different, usually related, species. Genes commonly flow back and forth among plants via transfers of pollen.

GENES The smallest segment of DNA containing a hereditary message. Some are "jumping" genes that regulate activity of other genes.

GENE SPLICING The creation of novel genetic combinations by intentionally interspersing a new gene sequence into an existing genome.

GENETIC CODE The sequence of nucleotides responsible for protein synthesis in cells.

GENETIC ENGINEERING Experimental and industrial technologies used to alter the genome of an organism so that it can produce more or different molecules than it is naturally programmed to make.

GENOME The full complement of genes carried by a given organism, and the genetic information contained in one complete set of chromosomes.

HERBICIDE A pesticide that usually affects only plants; a chemical with killing or growth inhibiting effects on plants.

HORMONE A substance secreted by certain cells that carry a signal to influence the activity of other cells and organs.

HYBRID The offspring of two parents differing from each other in one or more genes.

MASTITIS Inflammation of the udder of cows.

METABOLISM The process by which an organism processes food or other energy sources.

MONOCLONAL ANTIBODIES Identical antibodies cloned from a single source and targeted to deactivate specific antigens.

MONOCULTURE A growth or colony containing a single, pure genetic line of organisms; a genetically uniform line of plants.

MUTATION A change in the genome that can alter various cell functions and may be lethal.

NATURAL SELECTION The process by which interactions between an organism and its environment lead to a differential rate of reproduction among genetic types in a population. As a result, some genes increase in frequency in a population, while others decline. Natural selection is a primary factor in evolution.

NUCLEOTIDE The building block of the DNA molecule consisting of an organic base, a phosphate, and a sugar.

PATHOGEN An organism that causes disease in another organism. Bacteria, fungi, and viruses are among the disease-causing agents of plants and animals.

PHENOTYPE The observable characteristics of an organism.

PLASMID A circular molecule of DNA found in bacteria that is self-replicating and carries two or more genes.

POLLINATION The transfer of pollen from the male to the female part of a flower.

PROTEIN A molecule of linked amino acids. Proteins act as biochemical catalysts and as structural components of organisms.

rBGH Abbreviation for recombinant bovine growth hormone, a genetically engineered stimulant of milk production. Also referred to as BST (bovine somatotropin).

RECOMBINANT DNA A new combination of genes spliced together on a single piece of DNA.

RNA Ribonucleic acid, which transmits and translates DNA's genetic instructions.

ROUNDUP READY™ The brand name for crop seed genetically engineered to be resistant to the herbicidal effects of glyphosate (Roundup®).

TERATOGENIC Capable of producing birth defects or other reproductive harms.

TRANSGENE Gene from a dissimilar organism or an artificially constructed gene that is added to another organism using genesplicing techniques.

TRANSGENIC An adjective describing an organism that contains genes not native to its genetic makeup.

TRANSGENIC ORGANISM An animal or plant or microorganism that has been genetically engineered using gene-splicing methods so that it contains genetic material from at least one unrelated organism, for example, bacteria, viruses, animals, and other plants.

VIRAL COAT PROTEIN The outer envelope of protein enclosing the nucleic acid core of a virus.

VIRUS A small particle that reproduces only inside a living cell and consists of a nucleic acid core and a protein coat.

Selected Bibliography

Ahmed, M. "Biotechnology in the high school classroom." *American Biology Teacher* 58:178–80. 1996.

Ausubel, K. *Seeds of Change: The Living Treasure*. San Francisco: HarperCollins, 1994.

Baines, W. *Biotechnology from A to Z*. New York: Oxford University Press, 1993.

Bartley, D. M., and Hallerman, E. M. "A global perspective on the utilization of genetically modified organisms in aquaculture and fisheries." *Aquaculture* 137:1–7. 1995.

Biotechnology and Sustainable Agriculture: A Bibliography. (SRB-94-13) Beltsville, MD: USDA National Agricultural Library, 1994.

Bud, R. *The Uses of Life: A History of Biotechnology*. Cambridge, England: Cambridge University Press, 1993.

Busch, L.; Lacy, W. B.; Burkhardt, J.; et al. *Plants, Power and Profit: Social, Economic, and Ethical Consequences of the New Biotechnologies*. Cambridge, MA: Basil Blackwell, 1991.

Campbell, K. H. S.; McWhir, J.; Ritchie, W. A.; and Wilmut, I. "Sheep cloned by nuclear transfer from a cultured cell line." *Nature* 380:64–66. 1996.

Carroll, W. L. "Introduction to recombinant DNA technology." *American Journal of Clinical Nutrition* 58: Suppl:249S–258S. 1993.

Coghian, A. "Scorpion gene virus on trial in Oxford." *New Scientist*, p. 3. December, 1994.

Cohen, R. "The snark was a boojum." *The Story of Biotechnology Gone Crazy with Our Milk Supply and the Proof That Bovine Hormones Regulate Cancer in Humans*. Privately published. 1994. PO Box 36, Oradell, NJ 07649.

Collins, G. B., and Petolino, J. G. *Applications of Genetic Engineering to Crop Improvement*. Dordrecht, The Netherlands: Martinus Nijhoff/Dr. W. Junk Publishers. 1984.

Crouch, M. "Biotechnology is not compatible with sustainable agriculture." *Journal of Agricultural and Environmental Ethics* 8:98–111. 1995.

Dawkins, K. *Gene Wars: The Politics of Biotechnology*. New York: Seven Stories Press, 1998.

Doyle, J., et al. "Effects of genetically engineered microorganisms on microbial population and processes in natural habitats." *Advances in Applied Microbiology* 40:237–41, 1995.

Engel, K. H.; Takeoka, G. R.; and Teranishi, R. *Genetically Modified Foods: Safety Issues*. American Chemical Society Symposium Series 605. 1995.

Epstein, S. S. "Unlabelled milk from cows treated with biosynthetic growth hormones: A case of regulatory abdication." *International Journal of Health Services* 26:173–185. 1996.

Fano, A., et al. *Of Pigs, Primates and Plagues: A Layperson's Guide to the Problems with Animal-to-Human Organ Transplants*. New York: Medical Research Modernization Committee. Undated.

Food and Drug Administration, USDA. Conference on scientific issues related to potential allergenicity in transgenic foods crops, Annapolis, MD (Docket 3 94N-0053). Baltimore, MD: Free State Reporting, 1994.

Food and Drug Administration, USDA. "Food labeling: Foods derived from new plant varieties." *Federal Register* 58(80):25837–41. 1993.

Food and Drug Administration, USDA. "Statement of policy: Foods derived from new plant varieties." *Federal Register* 57 (104):22984–3004. 1992.

Food of the Future: The Risks and Realities of Biotechnology. Conference Proceedings, November 16–17, 1995, Oegstgeest, The Netherlands. Consumers International, 24 Highbury Crescent, London, England N5 1RX.

Food Quality and Safety: Innovative Strategies May Be Needed to Regulate New Food Technologies. Washington, DC: General Accounting Office (GAO/RCED-93-142), 1993.

Fox, M. W. *Eating with Conscience: The Bioethics of Food.* Troutdale, OR: NewSage Press, 1997.

Fox, M. W. *Superpigs and Wondercorn: The Brave New World of Biotechnology and Where It All May Lead.* New York: Lyons & Burford, 1992.

Gaull, G. E., and Goldberg, R. A., eds. *New Technologies and the Future of Food and Nutrition.* New York: John Wiley, 1991.

Goldburg, R., et al. *Biotechnology's Bitter Harvest: Herbicide-Tolerant Crops and the Threat to Sustainable Agriculture.* Washington, DC: Biotechnology Working Group, 1990.

Goodman, D.; Sorj, B.; and Wilkinson, J. *From Farming to Biotechnology: A Theory of Agro-industrial Development.* Oxford: Blackwell Publishers, 1987.

Hackett, P. B. "Genetic engineering of Minnesota superfish." *SSAS Bulletin: Biochem and Biotech* 9:69–76. 1996.

Hallberg, M., ed. *Bovine Somatotropin and Emerging Issues: An Assessment.* Boulder, CO: Westview Press, 1992.

Hettinger, N. "Patenting life: Biotechnology, intellectual property, and environmental ethics." *Environmental Affairs Law Review* (Boston College) 22:267–305. 1995.

Ho, Mae-Wan. *Genetic Engineering: Dream or Nightmare?* Bath, England: Gateway Books, 1998.

Hoban, T. J., and Kendall, P.A. *Consumer Attitudes About Food Biotechnology: Project Summary*. USDA Extension Service, North Carolina State University, Raleigh, NC. 1993.

Hobbelink, H. *Biotechnology and the Future of World Agriculture*. London: Zed Books, 1991.

Holdrege, C. *Genetics and the Manipulation of Life: The Forgotten Factor of Context*. Hudson, NY: Lindisfarne Press, 1996.

Holland, A., and Johnson, A., eds. *Animal Biotechnology and Ethics*. London: Chapman and Hall, 1998.

Holmes, M. T., and Ingham, E. R. "The effects of genetically engineered microbes on soil food webs." Abstract. *Ecological Society of America* 25(2). 1994. See also Holmes, M. T. et al., "Effects of *Klebstella planticola* on soil biota and wheat growth in sandy soil," *Applied Soil Ecology* 326: 1–12. 1998.

Hopkins, D. D., and Goldberg, R. A., eds. *A Mutable Feast: Assuring Food Safety in the Era of Genetic Engineering*. New York: Environmental Defense Fund, 1991.

Hubbard, R., and Wald, E. *Exploding the Gene Myth*. Boston: Beacon Press, 1993.

Jordan, J. P. "Biotechnology: Promise or pitfall." In: *Forest and Crop Biotechnology: Progress and Prospects*. Valentine, F. A., ed. New York: Springer-Verlag, 1988, pp. 1–10.

Juma, C. *The Gene Hunters: Biotechnology and the Scramble for Seeds*. Princeton, NJ: Princeton University Press, 1989.

Kapuscinski, A. R., and Hallerman, E. M. "Transgenic fishes." American Fisheries Society Position Statement. *Fisheries* 15:2–4. 1990.

Kay, L. E. *The Molecular Vision of Life: Caltech, the Rockefeller Foundation and the Rise of the New Biology*. Oxford, England: Oxford University Press, 1993.

Kenney, M. *Biotechnology: The University-Industrial Complex*. New Haven, CT: Yale University Press, 1986.

Kevles, D. J., and Hood, L., eds. *The Code of Codes: Scientific and Social Issues of the Human Genome Project.* Cambridge, MA: Harvard University Press, 1992.

Khor, M. "Genetic engineering: Science, ecology and policy." *Third World Resurgence* 65/66:22–29. 1995. Third World Network, 228 Macalister Road, 10400 Penang, Malaysia.

Kimbrell, A. *The Human Body Shop.* New York: Harper Collins, 1993.

Kloppenburg, J. *First the Seed: The Political Economy of Plant Biotechnology.* New York: Cambridge University Press, 1988, pp. 1492–2000.

Kloppenburg, J. Jr., et al. *Does Technology Know Where It's Going?* Edmonds, WA: Edmonds Institute, 1996.

Krimsky, S., and Wrubel, R. *Agricultural Biotechnology and the Environment.* Chicago: University of Illinois Press, 1996.

Lappe, M., and Bailey, B. *Against the Grain: Biotechnology and the Corporate Takeover of Your Food.* Monroe, ME: Common Courage Press, 1998.

Leowontin, R. *The Doctrine of DNA.* New York: Penguin Books, 1993.

Lesney, M. S., and Smocovitis, V. "Assessing the human genome project: Effects on world agriculture." *Agriculture and Human Values* XI (1):10-18. 1994.

Maramorosch, K., ed. *Biotechnology for Biological Control of Pests and Vectors.* Boca Raton, LA: CRC Press, 1991.

McCullum, C. *The New Biotechnology Era: Issues for Nutritional Policy.* Washington, DC: Community Nutrition Institute, 1995.

Medical Research Modernization Committee. *Letter to the FDA Regarding Labelling of Genetically Engineered Food Products.* 1997. PO Box 2751, Grand Central Station, New York, NY 10163-2751.

Mellon, M., and Rissler, J., eds. *Now or Never: Serious New Plans to Save a Natural Pest Control.* Cambridge, MA: Union of Concerned Scientists, 1998.

Mooney, P. "The parts of life: Agricultural biodiversity, indigenous knowledge, and the role of the third system." *Development Dialogue*. Special Issue. Uppsala, Sweden: Dag Hammarskjöld Centre, 1998.

Mooney, P., and Fowler, C. *Shattering: Food, Politics and the Loss of Genetic Diversity*. Tucson: University of Arizona Press, 1990.

Nelkin, D., and Lindee, M. S. *The DNA Mystique: The Gene as a Cultural Icon*. San Francisco: W.H. Freeman, 1995.

Nordlee, J. A., et al. "Identification of a Brazil nut allergen in transgenic soybeans." *New England Journal of Medicine* 334:688–92. 1996.

O'Brien, T. *Gene Transfer and the Welfare of Farm Animals*. Compassion in World Farming Trust, 5a Charles Street, Petersfield, England GU32 3EH. 1995.

Office of Technology Assessment. *Genetic Technology: A New Frontier*. Boulder, CO: Westview Press, 1982.

Patience, C., et al. "Infection of human cells by an endogenous retrovirus of pigs." *Nature Medicine* 3:282–286. 1997.

Raeburn, P. *The Last Harvest: The Genetic Gamble That Threatens to Destroy American Agriculture*. New York: Simon & Schuster, 1995.

Rissler, J., and Mellon, M. *The Ecological Risks of Engineered Crops*. Cambridge, MA: MIT Press, 1996.

Rissler, J., and Mellon, M. *Perils Amidst the Promise: Ecological Risks of Transgenic Crops in a Global Market*. Cambridge, MA: Union of Concerned Scientists, 1991.

Rollin, B. *The Frankenstein Syndrome*. Cambridge, England: Cambridge University Press, 1995.

Schmidt, K. "Genetic engineering yields first pest-resistant seeds." *Science* 265:739. 1994.

Shiva, V. *Biopiracy: The Plunder of Nature and Knowledge*. Boston: South End Press, 1997.

Shiva, V. *Monocultures of the Mind: Perspectives on Biodiversity and Biotechnology*. London: Zed Books, 1993.

Suzuki, D. T., and Knudtson, P. *Genethics: The Clash Between the New Genetics and Human Values*. Cambridge, MA: Harvard University Press, 1990.

Thompson, P. B. *Food Biotechnology in Ethical Perspective*. London: Chapman Hall, 1997.

Traavik T., *Too early may be too late: Ecological risks associated with the use of naked DNA as a biological tool for research, production, and therapy*. Research report for DN, Directorate for Nature Management, Trondheim, Norway, 1999.

Turner, J. D.; Delaquis, A.; and Karatzas, C. N. "Dietary needs of transgenic animals need considering." *Feedstuffs*, pp. 12–14. January 15, 1996.

Villalobos, V. M. "Impacts of biotechnology in international agriculture and forestry." Chapter 17 in B.W.S. Sobral, ed., *The Impact of Plant Molecular Genetics*. Boston: Birkhauser, 1990.

Ward, P. P., et al. "A system for production of commercial quantities of human lactoferrin: A broad spectrum natural antibiotic." *Biotechnology* 13:498–503. 1995.

Wilkie, T. *Perilous Knowledge: The Human Genome Project and Its Implications*. Berkeley: University of California Press, 1993.

Wirz, J., and Van Buren, E., eds. *The Future of DNA*. Dordrecht, The Netherlands: Kluwer Academic Publishers, 1997.

World Bank. *Agricultural Biotechnology: The Next Green Revolution?* Washington, DC: World Bank, 1990.

Resources

ORGANIZATIONAL RESOURCES AND INFORMATION

AgBiotech Reporter. AgBiotechnology Research/Business News, PO Box 7, Cedar Falls, IA 50613.

Biotech Reporter. Biotechnology Research/Business News, PO Box 7, Cedar Falls, IA 50613.

Biotechnology and Development Monitor. A free, quarterly, joint publication of the Department of Political Science of the University of Amsterdam and the Netherlands Ministry of Foreign Affairs Special Programme, Biotechnology and Development Cooperation, University of Amsterdam, Oudezids Achterburgwal 273, 1012 DL Amsterdam, The Netherlands.

Biotechnology Information Center, National Agricultural Library, 10301 Baltimore Boulevard, Beltsville, MD 20705-2351.

Center for Biotechnology Policy and Ethics, Texas A&M University, 329 Dulie Bell Building, College Station, TX 77843-4355.

Community Nutrition Institute, 910 17th Street NW, No. 413, Washington, DC 20006.

Council for Responsible Genetics Newsletter, 5 Upland Road, Suite 3, Cambridge, MA 02140.

Eubios Journal of Asia and International Bioethics. Eubios Ethics Institute, 31 Colwyn Street, Christchurch 5, New Zealand.

The Gene Exchange: A Public Voice on Biotechnology and Agriculture. Journal of the Union of Concerned Scientists, 1616 P Street NW, Washington, DC 20036.

The Loka Institute, PO Box 355, Amherst, MA 01004-0355. A non-profit organization dedicated to making science and technology responsive to democratically dedicated social and environmental concerns.

Mouse Genome. Quarterly journal reviewing mouse genetic studies of Oxford University Press, Oxford, England.

National Agricultural Biotechnology Council, 159 Biotechnology Building, Cornell University, Ithaca, NY 14853-2703. Publishes annual symposia.

Nature Biotechnology. Monthly publication from Nature America, Inc., 345 Park Avenue South, New York, NY 10010-1707.

Pesticide Action Network North America, 116 New Montgomery, No. 810, San Francisco, CA 94105.

Risky Business: Biotechnology and Agriculture. 24-minute video. Bullfrog Films, PO Box 149, Oakley, PA 19547.

The Splice of Life Newsletter. The Genetics Forum, 5-11 Worship Street, Third floor, London, England EC2A 2HB.

WEB SITE RESOURCES

Center for Food Safety (Washington, DC): www.icta.org

Consumer Right to Know Campaign for Mandatory Labelling and Long-term Testing of All Genetically Engineered Foods: www.natural-law.ca/genetic/geindex.html

Cultural Survival: www.cs.org

ELSA on-line forum. Discusses ethical, legal and social aspects of biotechnology. Wageningen Projects, The Hague, The Netherlands: www.w-projects.net

Environmental Defense Fund: www.edf.org

Gene Letter. Contains information about social issues in genetics as well as many interesting links: www.geneletter.org

Genline: www.hslib.washington.edu/helix

Genome Database Homepage: www.gdb.org

Greenpeace Biodiversity Campaign: www.greenpeace.org/cbio.html

Institute for Agriculture and Trade Policy: www.iatp.org

The Loka Institute. A nonprofit organization dedicated to making science and technology responsive to democratically dedicated social and environmental concerns: www.amherst.edu/-loka.

National Center for Biotechnology: www.ncbi.nlm.nih.gov/

National Human Genome Research Institute: www.nhgri.nih.gov

Organic Consumer Alliance: www.organicconsumers.org

Pesticides Action Network: www.panna.org/panna/

Physicians and Scientists Against Genetically Engineered Food. An international network for a moratorium on commercial release of genetically engineered products. Web site: "Genetically Engineered Foods-Safety Problems," http://home1.swipnet.se/~w-18472/indexing.htm

The Pure Food Campaign: www.purefood.org

Union of Concerned Scientists: www.ucsusa.org

Index